GRAPHICAL
ENUMERATION

GRAPHICAL ENUMERATION

Frank Harary

UNIVERSITY OF MICHIGAN
ANN ARBOR

Edgar M. Palmer

MICHIGAN STATE UNIVERSITY
EAST LANSING

ACADEMIC PRESS   New York and London   1973

ACADEMIC PRESS, INC.
111 Fifth Avenue, New York, New York 10003

United Kingdom Edition published by
ACADEMIC PRESS, INC. (LONDON) LTD.
24/28 Oval Road, London NW1

LIBRARY OF CONGRESS CATALOG CARD NUMBER: 72-82653

AMS (MOS) 1970 Subject Classification: 05C30

PRINTED IN THE UNITED STATES OF AMERICA

How do I love thee?
Let me count the ways.

Elizabeth Barrett Browning

To Jayne and Jane

The Royal Mathematician was a bald-headed,
nearsighted man, with a skullcap on his head and a
pencil behind each ear. He wore a black suit with
white numbers on it.

"I don't want to hear a long list of all the things
you have figured out for me since 1907," the King
said to him. "I just want you to figure out right now
how to get the moon for the Princess Lenore. When
she gets the moon, she will be well again."

"I am glad you mentioned all the things I have
figured out for you since 1907," said the Royal
Mathematician. "It so happens that I have a list
of them with me."

James Thurber, *"Many Moons"*

CONTENTS

2 PÓLYA'S THEOREM

3 TREES

4 GRAPHS

5 DIGRAPHS

6 POWER GROUP ENUMERATION

APPENDIXES

BIBLIOGRAPHY

PREFACE

The first question asked by many students in a course in graph theory is "How many graphs are there?" This is also the first problem we attempted. As circumstances had it, we learned by a most circuitous procedure that George Pólya had already counted graphs with a given number of points and lines. Starting from his formulas, it was a relatively routine matter to enumerate rooted graphs, connected graphs, and directed graphs. Subsequently, we counted various other types of graphs and when we had temporarily exhausted all the easy counting problems, we published a paper presenting 27 unsolved enumeration problems. By now, almost half of these problems have been resolved, and successive revisions of the original list of 27 unsolved enumeration problems were prepared. Our closing chapter brings this topic up to date.

Although Euler counted certain types of triangulated polygons in the plane, the major activity in graphical enumeration was launched in the preceding century. Cayley counted three types of trees: labeled trees, rooted trees, and ordinary trees. Even earlier, the world's first electrical engineer,

Kirchhoff, implicitly had found the number of spanning trees in a given connected graph, and thus in particular, the number of labeled trees. In one of the earliest instances of support of combinatorial research by the military (aside from Archimedes), Major P. A. MacMahon wrote a comprehensive treatise that touched on graphical enumeration, but only peripherally. There is another pre-Pólya innovator in the art of combinatorial enumeration. This largely unsung hero, J. Howard Redfield, wrote exactly one paper on the subject; in it he anticipated many of the counting methods and results found subsequently. His paper went almost completely unrecognized. Long after Pólya's great work served as the impetus for most of the contemporary research on the counting of graphs, proper acknowledgment to Redfield was accorded.

Although we are restricting ourselves to the enumeration of various kinds of graphs, there are many types of configurations that can be so handled. The following structures, none of which is blatantly graphical at first blush, have all been enumerated by clever transformations into graphs or subgraphs: automata, finite topologies, boolean functions, necklaces, and chemical isomers.

It is not only true that a full book can be written on each of our ten chapters, but a fortiori, an entire book has been written on one of the sections of our first chapter: a formal but comprehensive monograph entitled "Counting Labeled Trees" by John Moon. Clearly the material to be included in each chapter must necessarily be a matter of personal taste.

The plan of the book is as follows. We begin with labeled graphs in Chapter 1, both in order to get them out of the way and because they are much easier to count. We then develop the basic enumeration theorem of Pólya in Chapter 2. With this available, we count in Chapter 3 an enormous variety of trees and then in Chapters 4 and 5 various kinds of graphs and digraphs. Chapter 6 presents the powerful Power Group Enumeration Theorem and shows how to apply it. Chapter 7, Superposition, counts those configurations that can be constructed by "plopping things on top of other things." Nonseparable graphs, also known as blocks, are then counted in Chapter 8 using the ingenious methods conceived by the hero of unsolved enumeration problems, R. W. Robinson. Some mathematicians feel that a knowledge of the order of magnitude of the number of configurations of a certain type is more important than the exact number in a form which is inconvenient for calculations. Rather than report lower and upper bounds, we develop exact asymptotic numbers in Chapter 9 for several different graphical structures. Necessarily this is only illustrative, as again a whole book can be written on graphical asymptotics. Finally as a special feature we conclude with a new comprehensive definitive list of unsolved graphical enumeration problems.

The exercises range widely in difficulty from routine to intractible. Thus not all the exercises are intended to be worked out in detail by the reader. Frequently, counting formulas are given in exercises in order to include this information in the book. There are also abundantly many exercises within the text, not labeled as such, in the form of results whose proofs are omitted. We have found it convenient to indicate Equation 7 of Section 1 of Chapter 3 by the ordered triple denoted (3.1.7) and trust that the reader will forgive us for using this complicated notation. The end of a proof is marked by the symbol //.

It is our hope and belief that the present volume will make enumeration techniques more available and more unified. In turn this should serve as a stimulus for the investigation of open counting questions.

Acknowledgments

We owe special thanks to the following typists of the Department of Mathematics at Michigan State University who were most courteous, cooperative, accurate, and rapid in the preparation of several drafts of this book: Frieda Martin, Glendora Milligan, Darlene Robel, Terri Shaull, Nancy Super, Kathy Trebilcott, Mary Trojanowicz, and especially Mary Reynolds.

We are grateful to several sources for financial support while we were engaged in these research activities. Briefly but pleasantly, a grant from the Office of Naval Research supported one of us when we worked together launching this work at University College London during 1966–1967, where C. A. Rogers was our genial host. More recently, we have been respectively supported during summer intervals by the Air Force Office of Scientific Research and by the National Science Foundation.

We thank all whose names appear in the references. Helpful comments were made at various times by B. Manvel, R. C. Read, P. K. Stockmeyer, R. W. Robinson, and A. J. Schwenk. Very special thanks are due to John Riordan who gave the entire book his meticulous attention and offered many helpful suggestions. Most emphatically, each of us also thanks the other.

Finally, we thank Academic Press for their enthusiastic and effective support of graph theory and combinatorial theory. Tangible evidence of this can be found in the books cited in the bibliography and also in the existence of the first journal devoted to this fascinating subject, the *Journal of Combinatorial Theory*, founded by F. Harary and G.-C. Rota.

We offer ten cents (one U.S. dime) for each first notification of a misprint sent to either of us. Unlike Gilbert and Sullivan, we intend to continue talking to each other. Unlike Allendoerfer and Oakley, we do not blame each other for the misprints, but we join in blaming the publisher.

Ann Arbor, Michigan

FRANK HARARY

East Lansing, Michigan

EDGAR M. PALMER

Don't rely too much on labels,
For too often they are fables.

C. H. Spurgeon

Chapter 1 | LABELED ENUMERATION

We consider labeled enumeration problems first because they always appear to be much easier to solve than the corresponding unlabeled problems. For example, the number of labeled graphs is instantly found from first principles, while the determination of the number of unlabeled graphs requires a considerable amount of combinatorial theory including Pólya's Theorem.

We shall present in this chapter a selected sample of some of the outstanding and interesting solutions to labeled enumeration problems in graph theory, including the determination of the number of labeled graphs, connected graphs, blocks, eulerian graphs, k-colored graphs, acyclic digraphs, trees, and eulerian trails in an eulerian digraph. Often several different solutions to the same problem will be provided so that the reader has an opportunity to become acquainted with a variety of useful tricks, skills, devices, and schemes. For example, we shall see that when dealing with labeled enumeration problems, the exponential generating functions provide a natural vehicle for carrying sufficient information for a solution. On the

1

other hand, by examining a small amount of data, one can often quickly find a required formula which can then be verified by an induction argument.

1.1 THE NUMBER OF WAYS TO LABEL A GRAPH

A *graph G of order p* consists of a finite nonempty set $V = V(G)$ of p *points* together with a specified set X of q unordered pairs of distinct points; this automatically excludes *loops* (lines joining a point to itself) and multiple lines (in parallel). A pair $x = \{u, v\}$ of points in X is called a *line* of G and x is said to *join u* and v. The points u and v are *adjacent*; u and x are *incident* with each other, as are v and x. A graph with p points and q lines is called a (p, q) *graph*. Our terminology will follow that in the book on graph theory [H1]. However, we plan to include most definitions.

It is most convenient and illuminating to represent graphs by diagrams. Consider the graph G chosen at random with

$$V = \{v_1, v_2, v_3, v_4\}$$

and

$$X = \{\{v_1, v_2\}, \{v_2, v_3\}, \{v_3, v_4\}, \{v_4, v_1\}, \{v_1, v_3\}\}.$$

This is illustrated by the diagram in Figure 1.1.1. Only the names of the points have been used in this diagram. The five lines of G are represented by the line segments which join the pairs of points in the figure. The diagrams of all graphs of order 4, arranged by number of lines, are shown in Figure 1.1.2. Henceforth we shall also refer to such diagrams as graphs by an abuse of language which will cause no confusion.

In a *labeled graph* of order p, the integers from 1 through p are assigned to its points. For example, the random graph (of Figure 1.1.1) can be labeled in the six different ways indicated in Figure 1.1.3.

Thus two labeled graphs G_1 and G_2 are considered the same and called *isomorphic* if and only if there is a 1–1 map from $V(G_1)$ onto $V(G_2)$ which preserves not only adjacency but also the labeling. One can easily see then, that *all* of the different labelings of the random graph are displayed in Figure 1.1.3.

Figure 1.1.1

The graph with four points and five lines.

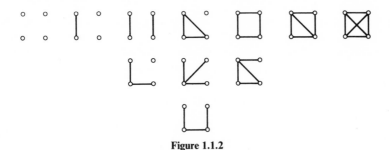

Figure 1.1.2

The 11 graphs of order 4.

Two natural questions now arise. The first asks: How many labeled graphs of order p are there? The second is: How many graphs of order p are there? The first question is so easy that we deal with it next. The second is much more difficult and will be treated in Chapter 4.

We shall answer the easier question by generalizing the problem ever so slightly to that of finding the number of labeled graphs with a given number of points *and* lines. Let $G_p(x)$ be that polynomial which has as the coefficient of x^k, the number of labeled graphs of order p which have exactly k lines. Such a polynomial is ordinarily called the "ordinary generating function" for labeled graphs with a given number of points and lines. If V is a set of p points, there are $\binom{p}{2}$ distinct unordered pairs of these points. In any labeled graph with point set V, each pair of points are either adjacent or not adjacent. The number of labeled graphs with precisely k lines is therefore $\binom{\binom{p}{2}}{k}$.

Theorem The ordinary generating function $G_p(x)$ for labeled graphs of order p is given by

$$G_p(x) = \sum_{k=0}^{m} \binom{m}{k} x^k = (1 + x)^m \qquad (1.1.1)$$

where $m = \binom{p}{2}$.

Since $G_p(x) = (1 + x)^m$ and the number G_p of labeled graphs of order p is $G_p(1)$, we see that

$$G_p = 2^{\binom{p}{2}}. \qquad (1.1.2)$$

Figure 1.1.3

The six different labelings of a graph.

Figure 1.1.4

The eight labeled graphs of order 3.

For $p = 3$; this formula is vividly illustrated in Figure 1.1.4. Thus there are eight labeled graphs of order 3 but only four graphs of order 3; and there are 64 labeled graphs of order 4, but only 11 graphs of order 4. The question then arises: In how many ways can a given graph be labeled? To provide an answer, we must consider the symmetries or automorphisms of a graph. A 1–1 map α from $V(G)$ to $V(G_1)$ that preserves adjacency is naturally called an *isomorphism*. If $G_1 = G$, then α is an *automorphism* of G. The collection of *all* automorphisms of G, denoted $\Gamma(G)$, constitutes a group called *the group of G*. Thus the elements of $\Gamma(G)$ are *permutations* acting on V. For example, the random graph G has exactly four automorphisms, so that $\Gamma(G)$ contains the permutations in the usual cyclic representation:

$$(v_1)(v_2)(v_3)(v_4), \qquad (v_1)(v_3)(v_2v_4), \qquad (v_1v_3)(v_2)(v_4), \qquad \text{and} \qquad (v_1v_3)(v_2v_4).$$

Let $s(G) = |\Gamma(G)|$, the order of the group G, denote the number of symmetries of G. Then the answer to the labeling problem posed above is provided in the following theorem.

Theorem   The number of ways of labeling a given graph G of order p is

$$l(G) = p!/s(G). \tag{1.1.3}$$

The proof is most easily obtained using some of the group theoretic results of Chapters 2 and 4, see [HPR1]. To illustrate, we simply observe that the random graph G has $p!/s(G) = 4!/4 = 6$ labelings, and the six different labeled graphs displayed in Figure 1.1.3 complete the verification of (1.1.3) for this graph G.

Although this theorem is stated only for graphs, similar versions of it hold for any finite structures with specified automorphism groups, such as rooted graphs, directed graphs, other relations of various types, simplicial complexes, functions, etc.

A *directed graph* or *digraph* D of order p consists of a finite nonempty set V of distinct objects called *points* together with a specified set X of q ordered pairs of distinct points of V. A pair $x = (u, v)$ of points in X is called an *arc* of D and u is said to be *adjacent to* v; u and x are *incident* with each other,

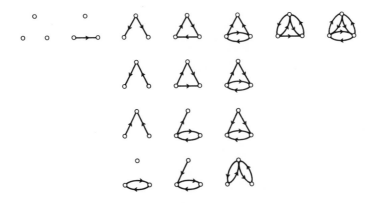

Figure 1.1.5

The 16 digraphs of order 3.

as are v and x. The *outdegree* of point u is the number of arcs with u as first point; the *indegree* as second point. The diagrams of all digraphs of order 3 are shown in Figure 1.1.5. As in the case of graphs, we refer to the diagrams themselves as digraphs.

Labeled digraphs of order p have the different integers 1 through p assigned to their points and *the group of a digraph D*, denoted $\Gamma(D)$, consists of the permutations of the points $V(D)$ of D that preserve adjacency. Since the number of labeled digraphs of order p with exactly k lines is $\binom{p(p-1)}{k}$, we have the following results which correspond to (1.1.1) and (1.1.2).

Theorem   The ordinary generating function $D_p(x)$ for labeled digraphs of order p is given by

$$D_p(x) = \sum_{k=0}^{p(p-1)} \binom{p(p-1)}{k} x^k = (1 + x)^{p(p-1)}. \tag{1.1.4}$$

Obviously $D_p(x) = G_p^2(x)$ so that

$$D_p(1) = 2^{p(p-1)} = G_p^2(1). \tag{1.1.5}$$

In a round-robin tournament, a given collection of players play a game in which the rules do not allow for a draw. Any two players encounter each other just once and exactly one emerges victorious. Therefore a *tournament* is a digraph in which every pair of points are joined by exactly one arc. We conclude this section by observing that the number of labeled tournaments of order p is precisely $2^{\binom{p}{2}}$, the number, as in (1.1.2), of labeled graphs of

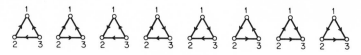

Figure 1.1.6

The eight labeled tournaments of order 3.

order p. This observation is verified for $p = 3$ by Figures 1.1.2 and 1.1.6. Furthermore, the natural correspondence between these two classes of graphs is indicated by the order in which they appear in the two figures. Each labeled tournament corresponds to that labeled graph in which the points with labels i and j are adjacent if and only if $i < j$ and the arc from i to j is present in the tournament.

1.2 CONNECTED GRAPHS

Let G be a graph and let $v_0, v_1, v_2, \ldots, v_n$ be a sequence of points of G such that v_i is adjacent to v_{i+1} for $i = 0$ to $n - 1$. Such a sequence together with these n lines, is called a *walk of length* n. If the lines $\{v_i, v_{i+1}\}$ for $i = 0$ to n are distinct, the walk is called a *trail*. If all the points are distinct (and hence the lines), it is called a *path of length* n. Then a *connected graph* is a graph in which any two points are joined by a path; see Figure 1.2.1. The number of labeled connected graphs of order 4 can be calculated by brute force if we apply (1.1.3) to each of the six graphs in Figure 1.2.1. The orders of the groups of these graphs, from left to right, are $2, 3, 2, 8, 4, 24$. Then from (1.1.3) it follows that the number of labeled, connected graphs of order 4 is 38. This information provides no hint as to how to determine a formula for C_p, the number of connected, labeled graphs of order p. To that end we require the next few definitions.

A *subgraph* H of a graph G has $V(H) \subset V(G)$ and $X(H) \subset X(G)$. A *component* of a graph is a maximal, connected subgraph. A *rooted graph* has one of its points, called the *root*, distinguished from the others. Two rooted graphs are *isomorphic* if there is a 1–1 function from the point set of one graph onto that of the other which preserves not only adjacency but also the roots. A similar requirement serves to describe rooted, labeled graphs. These ideas can now be used to obtain the following recursive formula.

Figure 1.2.1

The six connected graphs of order 4.

Theorem The number C_p of connected, labeled graphs satisfies

$$C_p = 2^{\binom{p}{2}} - \frac{1}{p} \sum_{k=1}^{p-1} k \binom{p}{k} 2^{\binom{p-k}{2}} C_k.$$ (1.2.1)

To prove (1.2.1) we observe that a different rooted, labeled graph is obtained when a labeled graph is rooted at each of its points. Hence the number of rooted, labeled graphs of order p is pG_p. The number of rooted, labeled graphs in which the root is in a component of exactly k points is $kC_k\binom{p}{k}G_{p-k}$. On summing from $k = 1$ to p, we arrive again at the number of rooted, labeled graphs, namely

$$\sum_{k=1}^{p} k \binom{p}{k} C_k G_{p-k}. \quad //$$

The values of C_p in Table 1.2.1 are listed in [S4].

TABLE 1.2.1

p	1	2	3	4	5	6	7	8	9
C_p	1	1	4	38	728	26 704	1 866 256	251 548 592	66 296 291 072

It is important to have at hand the concept of the *exponential generating function* and some of its associated properties. We shall therefore introduce these functions now and use them to provide an alternative form of (1.2.1).

For each $k = 1, 2, 3, \ldots$, let a_k be the number of ways of labeling all graphs of order k which have some property $P(a)$. Then the formal power series

$$a(x) = \sum_{k=1}^{\infty} a_k x^k / k!$$ (1.2.2)

is called the *exponential generating function* for the class of graphs at hand. Suppose also that

$$b(x) = \sum_{k=1}^{\infty} b_k x^k / k!$$ (1.2.3)

is another exponential generating function for a class of graphs with property $P(b)$.

The next lemma provides a useful interpretation of the coefficients of the product $a(x)b(x)$ of these two generating functions.

Labeled Counting Lemma The coefficient of $x^k/k!$ in $a(x)b(x)$ is the number of ordered pairs (G_1, G_2) of two disjoint graphs, where G_1 has property $P(a)$, G_2 has property $P(b)$, k is the number of points in $G_1 \cup G_2$ and the labels 1 through k have been distributed over $G_1 \cup G_2$.

To illustrate, let $C(x)$ be the exponential generating function for labeled, connected graphs,

$$C(x) = \sum_{k=1}^{\infty} C_k x^k/k!. \tag{1.2.4}$$

Then $C(x)C(x)$ is the generating function for ordered pairs of labeled, connected graphs. On dividing this series by 2, we have the generating function for labeled graphs which have exactly two components. Similarly $C^n(x)/n!$ has as the coefficient of $x^k/k!$, the number of labeled graphs of order k with exactly n components. If we let $G(x)$ be the exponential generating function for labeled graphs, we then have

$$G(x) = \sum_{n=1}^{\infty} C^n(x)/n!. \tag{1.2.5}$$

Thus we have the following exponential relationship for $G(x)$ and $C(x)$ found by Riddell [R14].

Theorem The exponential generating functions $G(x)$ and $C(x)$ for labeled graphs and labeled connected graphs come to terms in the following relation

$$1 + G(x) = e^{C(x)}. \tag{1.2.6}$$

Note that (1.2.6) remains true for multigraphs (Gilbert, [G2]). Riordan noticed the remarkable coincidence that $C_p = J_p(2)$, where $J_p(x)$ is the enumerator of trees by number of inversions [MR1] and thus obtained the following recurrence for C_p:

$$C_p = \sum_{k=1}^{p-1} \binom{p-2}{k-1} (2^k - 1) C_k C_{p-k}. \tag{1.2.7}$$

Furthermore, it is evident that if the exponential generating function for a class of graphs is known, then the exponential generating function for the corresponding connected graphs will be the formal logarithm of the first series, just as in (1.2.6) for all graphs.

Therefore we can state the following general result.

Corollary If $\sum_{m=0}^{\infty} A_m x^m = \exp\{\sum_{m=1}^{\infty} a_m x^m\}$, then for $m \geq 1$

$$a_m = A_m - m^{-1}\left(\sum_{k=1}^{m-1} k a_k A_{m-k}\right). \tag{1.2.8}$$

1.3 BLOCKS

The *removal of a point v* from a graph G results in that subgraph $G - v$ of G consisting of all points of G except v, and all lines not incident with v. A *cutpoint* of a graph is one whose removal increases the number of components. A *block* or *nonseparable graph* is connected, nontrivial, and has no cutpoints. We shall follow the procedure of Riddell [R14] and Ford and Uhlenbeck [FU1] in establishing relations between the generating functions for labeled blocks and labeled connected graphs. The approach used here is successful only for the labeled case. We shall see in Chapter 8 that a far more complex method is required for the enumeration of unlabeled blocks.

Since we are dealing with a labeled problem, we shall use exponential generating functions. Let $B(x)$ denote the series for labeled blocks so that

$$B(x) = \sum_{p=2}^{\infty} B_p x^p / p! \tag{1.3.1}$$

where B_p is the number of blocks with p points. It follows from formula (1.1.3) of the theorem on the number of ways to label a graph that the coefficient of x^p in $B(x)$ is the sum of the reciprocals of the orders of the groups of the (unlabeled) blocks with p points. Therefore from Figure 1.3.1 in which the small blocks are displayed together with their group orders, we have the first few terms of $B(x)$:

$$B(x) = \frac{1}{2}x^2 + \frac{1}{6}x^3 + \frac{5}{12}x^4 + \cdots = 1\frac{x^2}{2!} + 1\frac{x^3}{3!} + 10\frac{x^4}{4!} + \cdots. \tag{1.3.2}$$

Our aim is to prove the following theorem where $C'(x)$ and $B'(x)$ denote the usual formal derivative.

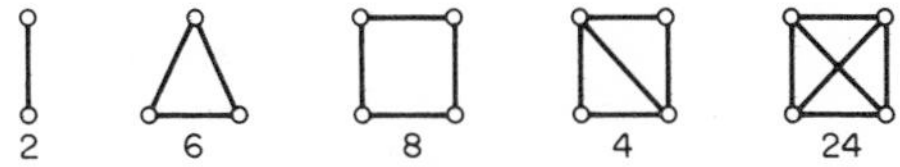

Figure 1.3.1

The small blocks and their symmetry numbers.

Theorem   The exponential generating functions $B(x)$ and $C(x)$ for labeled blocks and connected graphs are related by

$$\log C'(x) = B'(xC'(x)). \tag{1.3.3}$$

To verify this identity, let $R(x)$ be the exponential generating function for *rooted, connected, labeled graphs*, so that the coefficient of x^p in $R(x)$ is $R_p/p!$. Since $R_p = pC_p$ for all p we have

$$R(x) = x\, dC(x)/dx. \tag{1.3.4}$$

We denote by $R_n(x)$ the exponential series for rooted, connected, labeled graphs in which exactly n blocks contain the root. Thus $R_0(x) = x$ and

$$R(x) = \sum_{n=0}^{\infty} R_n(x). \tag{1.3.5}$$

Furthermore, $R_1(x)$ enumerates rooted, connected, labeled graphs with exactly one block incident with the root. Suppose $S(x)$ is the corresponding series in which the root is unlabeled; that is, the coefficient of $x^p/p!$ is the number of rooted, connected graphs with $p + 1$ points but the root is unlabeled. Then it follows from the Labeled Counting Lemma that $R_1(x) = xS(x)$, and hence $S(x) = R_1(x)/x$.

Again by the lemma, $(R_1(x)/x)^n/n!$ enumerates n-sets of such graphs where each root is unlabeled. If these n roots are identified and a single label is introduced for them, we shall have enumerated rooted, labeled, connected graphs with exactly n blocks at the root. Restoration of the labeled root is accomplished simply by multiplying by x:

$$R_n(x) = x(R_1(x)/x)^n/n!. \tag{1.3.6}$$

Combining the last two formulas yields

$$R(x) = x \exp(R_1(x)/x). \tag{1.3.7}$$

We now seek to express $R_1(x)$ in terms of $B(x)$ and $R(x)$. Observe that $(R(x)/x)^{k-1}$ counts $(k - 1)$-tuples of rooted, labeled, connected graphs in which the $k - 1$ roots are neither labeled nor included in the point count. That is, the coefficient of $x^p/p!$ in this series is the number of $(k - 1)$-tuples of such graphs with $p + k - 1$ points including the $k - 1$ roots and no labels on the roots. If we multiply this series by kB_k, we have counted rooted, connected graphs with one block at the root and in which only the labels 1 through k have been used for this block. Finally, to scatter all the labels we need only multiply by $x^k/k!$. Hence $xB_k(R(x))^{k-1}/(k - 1)!$ counts rooted, labeled, connected graphs with exactly one block of order k at the root.

Adding we have

$$R_1(x) = x \sum_{k=2}^{\infty} B_k(R(x))^{k-1}/(k-1)!. \qquad (1.3.8)$$

On combining the last two formulas which both involve $R_1(x)$ the result is

$$\log(R(x)/x) = \sum_{k=2}^{\infty} B_k(R(x))^{k-1}/(k-1)!. \qquad (1.3.9)$$

The proof is completed on substitution of $R(x) = xC'(x)$ from (1.3.4) in (1.3.9).   //

By comparing coefficients of x^p on each side of (1.3.3), we can arrive at a recursive formula for B_p. The coefficient of x^p on the left side of (1.3.3) can be expressed in terms of the coefficients of $C(x)$ using (1.2.8). For convenience let $h(p, k)$ denote the coefficient of x^p in $(xC'(x))^k$ so that the coefficient of x^p in the right side of (1.3.3) is

$$\sum_{k=2}^{p} B_k h(p, k-1)/(k-1)!. \qquad (1.3.10)$$

Hence the number of labeled blocks B_p can be expressed in terms of the numbers C_p of labeled connected graphs using (1.3.3). The method described here can be extended to include the number of lines as a second parameter without much more difficulty.

1.4 EULERIAN GRAPHS

In this section we shall derive, following the procedure of Read [R5] rather closely, the generating function for labeled eulerian graphs. The *degree* of a point v in a graph G is the number, denoted *deg v*, of lines of G which are incident with v. If every point of G has even degree, G is called *even*. An *eulerian graph* is a connected, even graph.

Let W_p be the number of labeled, even graphs of order p. Then the following rather surprising result occurs.

Theorem   The number of labeled, even graphs of order p equals the number of labeled graphs of order $p - 1$:

$$W_p = 2^{\binom{p-1}{2}}. \qquad (1.4.1)$$

To prove this result we now establish a 1–1 correspondence between these two classes of graphs. Consider any labeled graph G of order $p - 1$. Now G must have an even number of points of odd degree. Next we add to G a new point v which is assigned the label p. Finally, we construct a graph G' from G and v by specifying that v is adjacent to each of the points of G which has odd degree. This graph G' is a labeled even graph of order p. It is easily seen that this correspondence is 1–1, and that every labeled even graph of order p can be obtained in this way from some labeled graph of order $p - 1$.

$$//$$

We shall use generating functions to obtain a formula for the number of labeled eulerian graphs. Therefore let $W(x)$ be the exponential generating function for labeled even graphs, so that

$$W(x) = \sum_{p=1}^{\infty} 2^{\binom{p-1}{2}} x^p / p!. \tag{1.4.2}$$

Next, let U_p be the number of labeled, eulerian graphs of order p so that

$$U(x) = \sum_{p=1}^{\infty} U_p x^p / p! \tag{1.4.3}$$

is the corresponding exponential generating function.

Theorem   The exponential generating function $U(x)$ for labeled eulerian graphs satisfies

$$U(x) = \log(W(x) + 1), \tag{1.4.4}$$

and

$$U_p = 2^{\binom{p-1}{2}} - \frac{1}{p} \sum_{k=1}^{p-1} k \binom{p}{k} 2^{\binom{p-k-1}{2}} U_k. \tag{1.4.5}$$

Formula (1.4.4) follows from the fact mentioned after (1.2.6) that if the generating function for any class of graphs is known, then the generating function for the corresponding connected graphs is obtained by taking the formal logarithm of the first series. The recurrence relation (1.4.5) for U_p is a consequence of (1.4.4) and (1.2.8). $//$

For the first few terms of $U(x)$, we have

$$U(x) = x + \frac{x^3}{3!} + \frac{3x^4}{4!} + \frac{38x^5}{5!} + \cdots. \tag{1.4.6}$$

10   8   12   120

Figure 1.4.1

The four eulerian graphs of order 5.

The four eulerian graphs of order 5 are shown in Figure 1.4.1 together with the orders of their respective groups. According to (1.1.3) the reciprocals of these numbers should sum to $38/5!$, which is the coefficient of x^5 in $U(x)$, and indeed they do.

Next we consider the more difficult problem of determining the number of labeled eulerian graphs with a given number of points and lines. We seek to establish the following result of Read [R5].

Theorem The polynomial $w_p(x)$, which has as the coefficient of x^q the number of labeled graphs of even degree with p points and q lines, is given by

$$w_p(x) = \frac{1}{2^p}(1 + x)^{\binom{p}{2}} \sum_{n=0}^{p} \binom{p}{n}\left(\frac{1 - x}{1 + x}\right)^{n(p - n)}. \qquad (1.4.7)$$

For small p, we find that

$$w_1(x) = w_2(x) = 1, \qquad w_3(x) = 1 + x^3, \qquad \text{and} \qquad w_4(x) = 1 + 4x^3 + 3x^4.$$

Proof Let L be the set of all labeled graphs of order p with exactly q lines. Consider any graph G in L and arbitrarily multiply each of the labels 1 through p by $+1$ or -1. Since the labels will be positive or negative, each point can be referred to as "positive" or "negative" depending on the sign of its label. The numbers $+1$ or -1 are then assigned to each line as the product of the signs of its points. The *sign* of G, denoted $\sigma(G)$, is then defined as the product of the signs of its lines. There are, of course, 2^p ways in which the signs can be assigned to the labels of a given graph. On the other hand, suppose they have been allocated to the p integers which serve as labels; then there are $\binom{\binom{p}{2}}{q}$ different graphs with q lines and signed points determined by the given allocation of signs to the labels. These concepts are illustrated in Figure 1.4.2.

Since $\sigma(G)$ is the sign of the product of positive or negative numbers assigned to adjacent points, the positive points can be eliminated from this product. Thus

$$\sigma(G) = (-1)^a, \qquad (1.4.8)$$

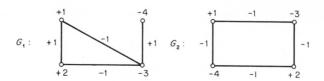

Figure 1.4.2

Two graphs with the same sign.

where a is the sum of the degrees of the negative points. On the other hand, obviously

$$\sigma(G) = (-1)^b, \tag{1.4.9}$$

where b is the number of negative lines of G, joining a negative point to a positive point.

Next we consider the sum $\sum \sigma(G)$ where the summation is for all labeled graphs in L *and* for the set S of 2^p possible allocations of $+1$ or -1 to the labels of the points. It follows from (1.4.8) and (1.4.9) that this sum can be written in two different ways:

$$\sum_{G \in L} \left\{ \sum_S (-1)^a \right\} = \sum_S \left\{ \sum_{G \in L} (-1)^b \right\}. \tag{1.4.10}$$

We first consider the left side of (1.4.10). If G is even, then a is even, whatever the allocations of signs in S. Hence $\sum(-1)^a = 2^p$ and G contributes 2^p to the left side of (1.4.11). If G is not even, at least one point v has odd degree. The allocations in S for which the label of v is positive and those for which it is negative are equinumerous and contribute opposite amounts to $\sum(-1)^a$. Hence G contributes nothing to the left side of (1.4.10). Thus the left side of (1.4.10) is 2^p times the number of even graphs in L.

Next we focus on the right side of (1.4.10) and consider an allocation in S for which n points are positive and $m = p - n$ are negative. There are $\binom{p}{n}$ such allocations. If there are k lines that join positive to negative points, these may occur in $\binom{nm}{k}$ different ways. The remaining $q - k$ lines can occur in

$$\binom{\binom{n}{2} + \binom{m}{2}}{q - k} \tag{1.4.11}$$

different ways. Summing from $k = 0$ to q, we obtain

$$\sum_{k=0}^q (-1)^k \binom{nm}{k} \binom{n(n-1)/2 + m(m-1)/2}{q-k} \tag{1.4.12}$$

as the contribution to the right side of (1.4.10) for each allocation with given n and m. This contribution is the coefficient of x^q in

$$(1 - x)^{nm}(1 + x)^{n(n-1)/2 + m(m-1)/2}. \tag{1.4.13}$$

Hence the right side of (1.4.10) is the coefficient of x^q in

$$\sum_{n=0}^{p} \binom{p}{n}(1 - x)^{nm}(1 + x)^{n(n-1)/2 + m(m-1)/2} \tag{1.4.14}$$

and this coefficient is 2^p times the number of even graphs in L. On observing that

$$\binom{n}{2} + \binom{m}{2} = \binom{p}{2} - n(p - n), \tag{1.4.15}$$

we obtain the final result that the required number of even graphs is the coefficient of x^q in the right side of (1.4.7). $\quad //$

We note in passing that the total number of labeled, even graphs is the number $w_p(1)$ obtained from (1.4.7) by setting $x = 1$ and observing the convention that $y^0 = 1$ even when $y = 0$:

$$w_p(1) = 2^{\binom{p-1}{2}}, \tag{1.4.16}$$

which verifies (1.4.1).

One can use (1.4.7) to obtain

$$w_5(x) = 1 + 10x^3 + 15x^4 + 12x^5 + 15x^6 + 10x^7 + x^{10}, \tag{1.4.17}$$

and the 64 labeled, even graphs counted by $w_5(x)$ can be obtained by labeling the seven even graphs displayed in Figure 1.4.3.

The exponential generating function $w(x, y)$ that enumerates all labeled, even graphs is

$$w(x, y) = \sum_{p=1}^{\infty} w_p(x)y^p/p!. \tag{1.4.18}$$

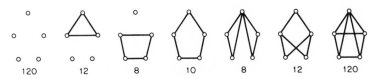

120   12   8   10   8   12   120

Figure 1.4.3

The even graphs of order 5 and their symmetry numbers.

To obtain the generating function $u(x, y)$ for labeled, eulerian graphs with a given number of points and lines, we need only take the logarithm of the series $1 + w(x, y)$:

$$u(x, y) = \log(1 + w(x, y)). \tag{1.4.19}$$

This observation follows from the two-variable version of the Labeled Counting Lemma.

1.5 THE NUMBER OF k-COLORED GRAPHS

A *colored graph* consists of a graph G with point set V, together with an equivalence relation on V such that no equivalent pair of points are adjacent. The k equivalence classes are regarded as the *colors* and G is called *k-colored*. Two *k-colored graphs are isomorphic* if there is a 1–1 correspondence between their point sets which preserves not only adjacency but also the colors. Note that the colors do not have fixed identities but are interchangeable. A given graph may be k-colored in many ways. For example, all the 3-colorings of a labeled graph of order 6 are shown in Figure 1.5.1 where the letters a, b, and c denote the colors and the integers denote the labels.

Following Read [R2], we shall find a formula for the number of labeled k-colored graphs of order p, generalizing a result of Gilbert [G2]. Let $p_1, \ldots,$ p_k be positive integers that form an ordered partition of p, so that

$$\sum_{i=1}^{k} p_i = p. \tag{1.5.1}$$

Writing $\{p\}$ for an arbitrary solution of (1.5.1), Read's formula takes the following form:

Theorem   The number $C_p(k)$ of k-colored labeled graphs of order p is

$$C_p(k) = \frac{1}{k!} \sum_{\{p\}} \binom{p}{p_1, \ldots, p_k} 2^{(p^2 - \Sigma p_i^2)/2}. \tag{1.5.2}$$

Figure 1.5.1

All four 3-colorings of a graph.

Proof Note that the number of k-colored, labeled graphs of order p in which the colors have fixed identities is $k! \, C_p(k)$. Hence we now consider the k colors as fixed. Each solution $\{p\}$ of (1.5.1) determines a k-part ordered partition of p, and so we seek the number of labeled graphs with p_i points of the ith color. The number of ways that the labels can be selected for the points is the multinomial coefficient

$$\binom{p}{p_1, p_2, \ldots, p_k}.$$

Obviously, there are

$$\binom{p}{2} - \sum_{i=1}^{k} \binom{p_i}{i} \tag{1.5.3}$$

pairs of points of different colors. Since each of these pairs may or may not be adjacent, we raise 2 to the power of (1.5.3), and use (1.5.1) to obtain for the total number of graphs with p_i points of color i, precisely the expression under the summation sign in (1.5.2). On summing over all solutions $\{p\}$ of (1.5.1), we have $k! \, C_p(k)$ and (1.5.2) is verified.   //

Note that the coefficient of x^q in

$$\frac{1}{k!} \sum_{\{p\}} \binom{p}{p_1, \ldots, p_k} (1 + x)^{(p^2 - \Sigma p_i^2)/2}$$

is the number of k-colored, labeled (p, q) graphs. For example, if we apply this assertion with $p = 4$, $q = 5$, and $k = 3$, we obtain six as the number of 3-colored labeled $(4, 5)$ graphs. This number six is also easily verified because there is only one unlabeled $(4, 5)$ graph and it can be labeled in six ways as in Figure 1.1.3.

A recursive formula for $C_p(k)$ is easily derived as a corollary:

$$C_p(k) = \frac{1}{k} \sum_{n=1}^{p-1} \binom{p}{n} 2^{n(p-n)} C_n(k - 1). \tag{1.5.4}$$

The verification of (1.5.4) can be accomplished by expressing the ordinary generating function for $C_p(k)$ in terms of that for $C_p(k - 1)$. The values of $k! \, C_p(k)$ for $p \leq 7$ are in Read [R2] and were used to derive Table 1.5.1.

Note that (1.2.8) cannot be used† to express the generating function for connected k-colored graphs in terms of that for k-colored graphs.

† Read wrote Wright that both Read [R2] and Wright [W3] were wrong. So Read and Wright wrote a joint erratum [RW1] to set things right. This may be wrong since Wright asserts that Wright wrote Read first.

TABLE 1.5.1

p/k	1	2	3	4	5	6	7
1	1	0	0	0	0	0	0
2	1	2	0	0	0	0	0
3	1	12	8	0	0	0	0
4	1	80	192	64	0	0	0
5	1	720	5 120	5 120	1 024	0	0
6	1	9 152	192 000	450 560	24 576	32 768	0
7	1	165 312	10 938 368	56 197 120	64 225 280	22 020 096	2 097 152

1.6 ACYCLIC DIGRAPHS

A *walk* of length n in a digraph D is determined by its sequence of points $v_0, v_1, \ldots, v_n$ in which v_i is adjacent *to* v_{i+1} for $i < n$. A *closed walk* has the same first and last points. A *cycle* is a nontrivial closed walk with all points distinct except the first and last. An *acyclic* digraph has no cycles. Labeled acyclic digraphs are now enumerated rather easily following Robinson [R20], but the unlabeled case requires more powerful machinery that is developed in Chapter 8.

A digraph E is *an extension* of D if D is the subgraph of E induced by the points of E with positive indegree. Every acyclic digraph must have at least one point of indegree zero [HNC1, p. 64]. Therefore every acyclic digraph with at least one arc is the extension of a unique proper subgraph. Furthermore, every acyclic digraph has many extensions, but each must be acyclic.

Suppose D is an acyclic digraph with exactly $n \geq 1$ points u_i of indegree zero and s other points v_i. We can form an extension E of D having exactly k points of indegree zero by adding k new points w_i and new arcs such that each of the n points u_i is adjacent from some new point w_i, but otherwise the transmitters w_i may be adjacent to any of the other points v_i of D. In Figure 1.6.1, the new points w_1, w_2, and w_3 were added and each old point u_1 and u_2 of indegree zero is adjacent from some w_i.

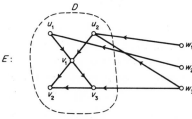

Figure 1.6.1

An extension of an acyclic digraph.

Thus all the acyclic digraphs of order p can be obtained by extending the acyclic digraphs of order less than p. Specifically, let a_p be the number of labeled acyclic digraphs of order p and let $a_{p,k}$ be the number of order p which have exactly $k \geq 1$ points of indegree zero. For $k = p$, we have of course $a_{p,p} = 1$, since the totally disconnected digraph is the only candidate. Clearly for all p,

$$a_p = \sum_{k=1}^{p} a_{p,k}. \qquad (1.6.1)$$

We shall now show how $a_{p,k}$ can be expressed in terms of $a_{p-k,n}$ with $n \leq p - k$. First we prove that the contribution to $a_{p,k}$ from all extensions of the $a_{p-k,n}$ digraphs with $p - k$ points of which exactly n have indegree zero is

$$(2^k - 1)^n 2^{k(p-n-k)} \binom{p}{k} a_{p-k,n}. \qquad (1.6.2)$$

We seek the number of labeled extensions E of the $a_{p-k,n}$ labeled acyclic digraphs D. For each of the $\binom{p}{k}$ ways of labeling the k new points w_i in E, there are $a_{p-k,n}$ labelings of the digraphs D to be extended. This accounts for the factor $\binom{p}{k} a_{p-k,n}$ in (1.6.2). Each of the n points of indegree zero in D must be adjacent *from* at least one of the k new points. Hence there are $2^k - 1$ possible combinations of arcs to each of these n points and therefore $(2^k - 1)^n$ for all of them. Each new point may or may not be adjacent to any of the $p - n - k$ points of positive indegree in D. Therefore there are 2^{p-n-k} possible combinations for each new point and hence $(2^{p-n-k})^k$ in all. On multiplying these factors, (1.6.2) is obtained.

Summing (1.6.2) over n, we have an expression for $a_{p,k}$.

Theorem   The number $a_{p,k}$ of labeled acyclic digraphs of order p which have exactly k points of indegree zero is

$$a_{p,k} = \sum_{n=1}^{p-k} (2^k - 1)^n 2^{k(p-n-k)} \binom{p}{k} a_{p-k,n}. \qquad (1.6.3)$$

Thus (1.6.1) and (1.6.3) can be used to determine a_p. These results can also be expressed in terms of generating functions. Let $a(x, y)$ have as the coefficient of $x^k y^{p-k}$ the number of labeled acyclic digraphs with p points, k of which have indegree zero. Then the first few terms of $a(x, y)$ are given by

$$a(x, y) = x + x^2 + 2xy + x^3 + 9x^2 y + 15xy^2 + x^4 + 28x^3 y$$
$$+ 198x^2 y^2 + 316xy^3 + x^5 + 75x^4 y + 1610x^3 y^2$$
$$+ 10710x^2 y^3 + 16885xy^4 + \cdots. \qquad (1.6.4)$$

Figure 1.6.2

The two acyclic digraphs of order 3 with two points of indegree zero.

For example, there are six ways in which that acyclic digraph D_1 of Figure 1.6.2 can be labeled and three in which the other, D_2, can be labeled. The total of nine corresponds to the term $9x^2y$ of $a(x, y)$.

1.7 TREES

A *tree* is a connected graph that has no cycles, see [H1, Chap. 4]. It is well known that every nontrivial tree has at least two endpoints (of degree 1). It follows that if T is a tree with p points and q lines then

$$q = p - 1. \tag{1.7.1}$$

All of the trees with as many as five points are shown in Figure 1.7.1 together with the number of ways in which each may be labeled. From these data, the number t_p of labeled trees with p points has its smallest values 1, 1, 3, 16, 125. Many authors have correctly surmised from this sequence that the counting formula is given by the next theorem.

Theorem   (Cayley)   The number t_p of labeled trees of order p is

$$t_p = p^{p-2}. \tag{1.7.2}$$

We shall sketch only the four proofs of Cayley, Prüfer, Pólya, and Kirchhoff, although there are many others no less interesting than these. A collec-

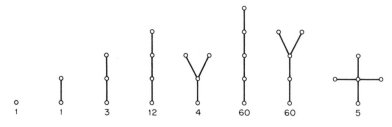

Figure 1.7.1

The trees of order up to 5, and the number of ways to label each.

tion of such proofs may be found in Moon's article [M1], as a sequel to which he wrote an entire book [M4] on the subject of counting various types of labeled trees.

Cayley [C2] suggested a correspondence between labeled trees and functions from a set of $p - 2$ objects to a set of p objects. For example, for $p = 5$ there are 5^3 functions from $\{a, b, c\}$ to $\{v_1, v_2, v_3, v_4, v_5\}$. These functions are enumerated by the polynomial

$$(v_1 + v_2 + v_3 + v_4 + v_5)^3. \tag{1.7.3}$$

Its terms correspond in a natural way to the functions. To illustrate, v_4^3 corresponds to the constant function $f(x) = v_4$, the term $3v_1v_3^2$ indicates the three functions that send a single element to v_1 while the other two go to v_3, and $6v_2v_3v_5$ gives the six functions that send one element to each of v_2, v_3 and v_5. Now if (1.7.3) is multiplied by $v_1v_2v_3v_4v_5$ to obtain

$$(v_1 + v_2 + v_3 + v_4 + v_5)^3 v_1v_2v_3v_4v_5, \tag{1.7.4}$$

then there is a correspondence between the terms in this product and the labeled trees of order 5. This correspondence is displayed in Figure 1.7.2 using the term $3v_1^2v_2v_3^3v_4v_5 = 3v_1v_3^2(v_1v_2v_3v_4v_5)$. Note that in the trees corresponding to $v_1^2v_2v_3^3v_4v_5$ the degree of the point labeled k is the exponent of v_k. This observation can be shown to be true in general and hence the number of labeled trees in which the point with label k has degree d_k is the multinomial coefficient

$$\binom{p - 2}{d_1 - 1, d_2 - 1, \ldots, d_p - 1}. \tag{1.7.5}$$

Cayley [C2] displayed this correspondence for $p = 6$ and dismissed the other cases with the remark that "It will be at once seen that the proof given for this particular case is applicable for any value whatever of p."

Prüfer [P10] obtained a correspondence between labeled trees of order p and $(p - 2)$-tuples $(a_1, a_2, \ldots, a_{p-2})$, where each a_k is an integer from 1 to p with repetitions permitted. Thus there are p^{p-2} such sequences. For a

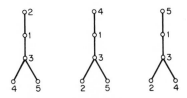

Figure 1.7.2

Labeled trees counted by $v_1^2v_2v_3^3v_4v_5$.

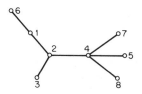

Figure 1.7.3

A labeled tree of order 8.

given labeled tree T, let v be the endpoint with the smallest label and let a_1 be the label of the point adjacent to v. Now to obtain a_2 repeat this step with $T - v$, the tree obtained from T by deleting v (and the line incident with v). The procedure is terminated when only two adjacent points remain. Note that the tree in Figure 1.7.3 corresponds to the sequence (2, 4, 1, 2, 4, 4).

Since each labeled tree of order p yields a unique $(p - 2)$-tuple, $t_p \leq p^{p-2}$.

To show that $t_p \geq p^{p-2}$, we describe a procedure for constructing a unique labeled tree from each $(p - 2)$-tuple, $(a_1, a_2, \ldots, a_{p-2})$. Following Moon [M4, p. 5], let b_1 be the smallest positive integer that does not occur in the $(p - 2)$-tuple and let $(c_2, \ldots, c_{p-2})$ denote the $(p - 3)$-tuple obtained from $(a_2, \ldots, a_{p-2})$ by diminishing all terms larger than b_1 by 1. Then $(c_2, \ldots, c_{p-2})$ consists of the numbers 1 through $p - 1$, and we can assume there is a corresponding tree T of order $p - 1$. Next relabel the points of T by adding 1 to each label that is larger than $b_1 - 1$. Then introduce a pth point labeled b_1 and join it to the point labeled a_1 in T. Thus a unique labeled tree is obtained which corresponds to the given $(p - 2)$-tuple.

Next we consider Pólya's method [P8] for determining the number of labeled trees. Since the number of rooted, labeled trees of order p is pt_p, the exponential generating function for these trees is given by

$$y = \sum_{p=1}^{\infty} pt_p x^p / p!. \tag{1.7.6}$$

Pólya found a functional equation for y and then applied Lagrange's inversion formula to determine t_p.

This functional equation for y is now derived. It follows from the Labeled Counting Lemma that $y^n/n!$ is the exponential generating function for n-sets of rooted labeled trees. These n-sets correspond precisely to rooted labeled trees in which the root has degree n but no label. More specifically, this correspondence is obtained by first adding a new point with no label to each n-set and by then joining this new point to each of the old roots. This idea is illustrated in Figure 1.7.4. Multiplication by x introduces a label for the

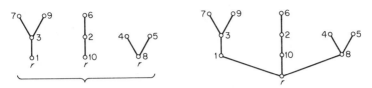

Figure 1.7.4

A 3-set of rooted trees and the corresponding tree whose root has degree 3.

new root and adds it to the point count. Thus $xy^n/n!$ enumerates rooted, labeled trees in which the root has degree n. On summing we obtain

$$y = \sum_{n=0}^{\infty} xy^n/n!, \tag{1.7.7}$$

and hence we arrive at the functional equation

$$y = xe^y. \tag{1.7.8}$$

To solve (1.7.8) for y in terms of x we apply the very useful special case of the formula of Lagrange given in Moon [M4, p. 26]; see also Pólya [P8].

Lagrange's Inversion Formula   If $\varphi(y)$ is analytic in a neighborhood of $y = 0$ with $\varphi(0) \neq 0$, then the equation

$$x = y/\varphi(y). \tag{1.7.9}$$

is uniquely solved by the generating function

$$y = \sum_{k=1}^{\infty} c_k x^k \tag{1.7.10}$$

whose coefficients are

$$c_k = (1/k!)\{(d/dy)^{k-1}(\varphi(y))^k\}_{y=0}. \tag{1.7.11}$$

On applying this inversion formula to (1.7.8) where $\varphi(y) = e^y$, we find that

$$y = \sum_{k=1}^{\infty} k^{k-1}x^k/k!, \tag{1.7.12}$$

and confronting this with (1.7.6), we again obtain the formula (1.7.2) for t_p.

To solve some labeled counting problems (see Exercises 1.13a, b and 1.14) it is convenient to use Lagrange's generalization of formula (1.7.11). In addition to the conditions on the function φ, we assume that $f(y)$ is another function analytic in a neighborhood of $y = 0$. Lagrange's general formula states that $f(y)$ can be expressed as a power series in x as follows:

$$f(y) = f(0) + \sum_{k=1}^{\infty} \frac{x^k}{k!} \left\{ \left(\frac{d}{dy}\right)^{k-1} [f'(y)\varphi^k(y)] \right\}_{y=0}. \qquad (1.7.13)$$

With $f(y) = y$, this formula implies (1.7.10) and (1.7.11). A proof of (1.7.13) can be found in Goursat and Hedrick [GH1].

A most interesting and useful result usually called the "Matrix-Tree Theorem" is implicit in the work of Kirchhoff [K3]. The number of labeled trees can be derived quickly as a corollary. The *adjacency matrix* $A = A(G) = [a_{ij}]$ *of a labeled graph* G of order p is the $p \times p$ matrix in which $a_{ij} = 1$ if v_i and v_j are adjacent and $a_{ij} = 0$ otherwise. Hence there is a 1–1 correspondence between labeled graphs of order p and $p \times p$ symmetric binary matrices with zero diagonal. Let $M(G)$ denote the matrix obtained from $-A$ by replacing the ith diagonal entry by deg v_i. A subgraph H of G *spans* G if every point of G is a point of H (see Figure 1.7.5.)

Matrix-Tree Theorem for Graphs For any connected labeled graph G, all cofactors of the matrix $M(G)$ are equal and their common value is the number of spanning trees of G.

The proof can be found in [H1, p. 152]. To illustrate, consider the graph G in Figure 1.7.5. It has three spanning trees since the 1, 4 cofactor, for example, of

$$M(G) = \begin{vmatrix} 2 & -1 & -1 & 0 \\ -1 & 2 & -1 & 0 \\ -1 & -1 & 3 & -1 \\ 0 & 0 & -1 & 1 \end{vmatrix} \quad \text{is} \quad - \begin{vmatrix} -1 & 2 & -1 \\ -1 & -1 & 3 \\ 0 & 0 & -1 \end{vmatrix} = 3.$$

The *complete graph* K_p with all lines present can be labeled in only one way and every spanning tree corresponds to a different labeled tree. Hence the

Figure 1.7.5

A graph and its three spanning trees.

number of labeled trees of order p is formed by applying the theorem to K_p. Each principal cofactor of $M(K_p)$ is the determinant of order $p - 1$:

$$\begin{vmatrix} p-1 & -1 & \cdots & -1 \\ -1 & p-1 & \cdots & -1 \\ \vdots & \vdots & \ddots & \vdots \\ -1 & -1 & \cdots & p-1 \end{vmatrix}.$$

On subtracting the first row from each of the others and adding the last $p - 2$ columns to the first, we arrive at an upper triangular matrix whose determinant is p^{p-2}.

1.8 EULERIAN TRAILS IN DIGRAPHS

We saw in Section 1.7 that the Matrix-Tree Theorem for Graphs provides one of several methods for counting labeled trees, by determining the number of spanning subtrees of a labeled K_p. We now develop only the statement of the extension of this theorem to digraphs, which gives the number of spanning subtrees of a given digraph D that are oriented toward each point. The object of this section is to apply this Matrix-Tree Theorem for Digraphs to a labeled eulerian digraph D in order to derive an explicit formula for the number of eulerian trails in D.

A *tree to a point* is obtained from a rooted tree T with root v by orienting all the arcs toward v. A *tree from a point* is the directional dual. Clearly these are both in 1–1 correspondence with rooted trees.

Consider the digraph D of Figure 1.8.1 whose points are labeled 1, 2, 3, 4, 5. There are just four spanning trees from point 1 and two spanning trees to point 1, as shown in Figure 1.8.1b and c.

Let D be a digraph with adjacency matrix A. Define the diagonal matrix M_{out} with i, i entry od v_i, the outdegree of v_i. Then let $C_{out} = M_{out} - A$. Thus every row sum in C_{out} is zero, but not necessarily every column sum. In fact the column sums of C_{out} are also all zero if and only if D is eulerian, as we shall soon see. Similarly define $C_{in} = M_{in} - A$. The important next result was found by Bott and Mayberry [BM2] and the proof is due to Tutte [T2].

(1.8.1) Matrix-Tree Theorem for Digraphs   All the cofactors of the ith row of C_{out} are equal, and their common value is the number of spanning trees of D to v_i. Dually, the common value of the cofactors of the ith column of C_{in} is the number of spanning trees from v_i.

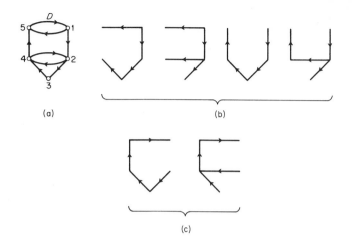

(a)   (b)

(c)

Figure 1.8.1

The spanning trees of D from and to the point labeled 1.

Although we omit the proof, this theorem is easily illustrated for the
digraph D of Figure 1.8.1, for which the two matrices of the theorem are

$$C_{\text{out}} = \begin{bmatrix} 2 & -1 & 0 & 0 & -1 \\ 0 & 2 & -1 & -1 & 0 \\ 0 & 0 & 1 & -1 & 0 \\ 0 & -1 & 0 & 2 & -1 \\ -1 & 0 & 0 & 0 & 1 \end{bmatrix},$$

$$C_{\text{in}} = \begin{bmatrix} 1 & -1 & 0 & 0 & -1 \\ 0 & 2 & -1 & -1 & 0 \\ 0 & 0 & 1 & -1 & 0 \\ 0 & -1 & 0 & 0 & -1 \\ -1 & 0 & 0 & 0 & 2 \end{bmatrix}.$$

From these it is verified at once from the first row of C_{out} and from the first
column of C_{in} that D has exactly four trees from point 1 and two trees to it,
as shown in Figure 1.8.1.

A *digraph* is called *eulerian* if there exists a closed spanning directed
walk passing through each arc exactly once. Such a walk is a directed *eulerian
trail*. One criterion for a digraph to be eulerian [H1, p. 204] is that it be con-

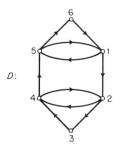

Figure 1.8.2

An eulerian digraph.

nected while each point has equal outdegree and indegree. For example, the digraph of Figure 1.8.1 is not eulerian, but D in Figure 1.8.2 is.

It follows from the definition of an eulerian digraph that C_{out} and C_{in} have the same diagonal, and are therefore equal. For the digraph D of Figure 1.8.2, this matrix is

$$
C = C_{out} = C_{in} = \begin{bmatrix}
2 & -1 & 0 & 0 & -1 & 0 \\
0 & 2 & -1 & -1 & 0 & 0 \\
0 & 0 & 1 & -1 & 0 & 0 \\
0 & -1 & 0 & 2 & -1 & 0 \\
-1 & 0 & 0 & 0 & 2 & -1 \\
-1 & 0 & 0 & 0 & 0 & 1
\end{bmatrix}.
$$

A first step in the proof of the Matrix-Tree Theorem for Graphs is the observation that in a matrix with all row sums and column sums zero, every cofactor has the same value. Therefore, by Theorem 1.8.1, every eulerian digraph has the same number of spanning trees to each point and from each point. For example, in the matrix above all cofactors equal 4, so there are four trees to each point as illustrated in Figure 1.8.3.

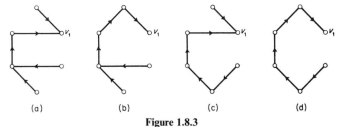

(a)   (b)   (c)   (d)

Figure 1.8.3

The spanning trees to v_1 in Figure 1.8.2.

We are now ready to apply the Matrix-Tree Theorem for Digraphs to the derivation of the number of eulerian trails in a given digraph. The proof follows the elegant exposition in Kasteleyn [K1]. The result itself was first found by van Aardenne–Ehrenfest and de Bruijn [BE1], although a special case appears earlier in Smith and Tutte [ST1]. Because each point v_i of an eulerian digraph D has od v_i = id v_i, we can denote this number by d_i.

Theorem   The number $e(D)$ of eulerian trails in a labeled eulerian digraph D in which c is the common value of the cofactors of $C = C_{out} = C_{in}$ is

$$e(D) = c \prod_i (d_i - 1)!. \tag{1.8.2}$$

Proof   Let v_1 be any point of an eulerian digraph D. We shall show that each eulerian trail E of D determines a unique spanning tree T to v_1, and that each such tree T determines exactly $\prod(d_i - 1)!$ eulerian trails. Since we have already seen that the number of spanning trees of D to each point is c, equation (1.8.2) will then be proved.

To construct the spanning tree to v_1 determined by a given eulerian trail E in a digraph D, call the *exit arc* from each point $v_i \neq v_1$ the last arc out of v_i when traversing E with starting (and finishing) point v_1. Thus only v_1 has no exit arc. Then define T as the spanning subgraph of D whose arcs are the exist arcs. Since in T, v_1 has outdegree 0 and all other points have outdegree 1, it must be a tree to v_1 by [HNC1, p. 283, Theorem 10.12].

Now let T be a fixed spanning tree to v_1 (among the c such trees). We proceed to construct all eulerian trails E associated with T in the manner of the preceding paragraph, that is, the exit arcs of E with respect to v_1 are the arcs of T. Since D is eulerian, we have already noted that od v_i = id v_i = d_i. In constructing E from T, one arc from each point $v_i \neq v_1$ is put aside for later use as the exit arc, and one arc from v_1 is reserved for use as the first arc of E. Then at each and every point v_i (including v_1), there are exactly $(d_i - 1)!$ orders in which the occurrences of arcs in E can appear. Since these occurrences are independent, we multiply these factorials to get the number of eulerian trails determined by T. But there are c such trees, proving (1.8.2).

                                                                        //

(1.8.3) **Corollary**   In an eulerian digraph, in which each $d_i = 1$ or 2, the number of eulerian trails equals the number c of spanning trees to each point.

This follows at once from the observation that every $(d_i - 1)! = 1$. We illustrate the corollary for D in Figure 1.8.2, in which each d_i is 1 or 2.

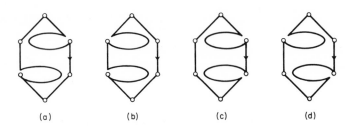

(a)   (b)   (c)   (d)

Figure 1.8.4

The eulerian trails of Figure 1.8.2.

Thus we know from the calculation of the cofactor preceding Figure 1.8.3 that D has exactly four eulerian trails. These are now displayed (Figure 1.8.4) in correspondence with the spanning trees of exit lines shown in the preceding figure.

EXERCISES†

1.1 Connected labeled digraphs.

1.2 Labeled orientations of a given (p, q) graph with symmetry number s.

1.3 Oriented labeled graphs and *signed* labeled graphs in which each line is positive or negative.

1.4 Labeled (p, q) graphs with no isolated points.   (Gilbert [G2])

1.5 (a) Labeled, connected, (p, q) graphs:

$$\sum_{k=1}^{p} \frac{(-1)^{k+1}}{k} \sum_{\{p\}} \frac{p!}{\prod p_i!} \binom{m}{k},$$

where $m = \sum_{i=1}^{k} \binom{p_i}{m}$ and the second sum is over all partitions $p_1 + \cdots + p_k = p$ of p with k parts.

(b) Labeled (p, q) blocks.   (Riddell [R14])

1.6 A labeled graph with point set $\{v_0, v_1, \ldots, v_{p-1}\}$ is a starred polygon if v_0 adj v_i implies that for all $k = 1$ to $p - 1$, v_k adj v_{k+i}, where the subscripts are taken mod p. There are $2^{(p-1)/2}$ starred polygons of order p. (Turner [T1])

---

† Whenever a class of graphs is mentioned in an exercise, it is understood that the reader is being asked to find a counting formula for them.

1.7   Labeled, even, general graphs (see Section 7.5).   (Read [R5])

1.8   Connected graphs with all points labeled except endpoints.

(Moon [M3])

1.9   Labeled trees in which each point has degree (a) 1 or 3, (b) degree 1 or n, where $n > 3$.

1.10   Labeled bicolored trees with m points of one color and n of the other:

$$n^{m-1}m^{n-1}.$$ (Scoins [S1])

1.11   Labeled homeomorphically irreducible trees:

$$(p-2)! \sum_{k=2}^{p} (-1)^{p-k} \binom{p}{k} \frac{k^{k-2}}{(k-2)!}.$$ (Read [R9])

1.12   Labeled trees with unlabeled endpoints. (*Hint:*   Use Stirling numbers of the second kind.)   (Harary, Mowshowitz, Riordan [HMR1])

1.13   (a)   Labeled 2-trees (see Section 3.5):

$$\binom{p}{2}(2p-3)^{p-4}.$$ (Beineke and Moon [BM1])

(b)   Labeled k-trees:

$$\binom{p}{k}(kp-k^2+1)^{p-+-2}.$$

(Beineke and Pippert [BP1])

(c)   Line-labeled 2-trees:

$$\frac{(2p-3)!}{2(p-2)!}(2p-3)^{p-4}.$$ (Palmer [P1])

1.14   A *plane graph* has been embedded in the plane so that no two lines intersect. Labeled, plane 2-trees:

$$p(p-1)^2 \frac{(5p-10)!}{(4p-6)!}.$$ (Palmer and Read [PR1])

1.15   (a)   Labeled, connected functional digraphs (see Section 3.4):

$$\sum_{k=1}^{p} \frac{p!}{(p-k)!}p^{p-k+1}.$$ (Rubin and Sitgreaves)

(b)   Labeled, connected unicyclic graphs.   (Moon [M4])

1.16 Labeled graphs with (a) no endpoints, (b) with a given number of endpoints. (Read [R9])

1.17 Labeled cacti (see Section 3.4) with c_2 lines in no cycle, c_3 triangles, c_4 quadrilaterals, . . . , c_n cycles of n points:

$$\frac{p!\,p^c}{\prod_{k=2}^{p}((k-1)!)^{c_k}c_k!},$$

where $c = \sum_{k=2}^{p} c_k - 2$. (Harary and Uhlenbeck [HU1])

1.18 (a) The number c_3 of triangles in a labeled graph is $\frac{1}{6}$ of the trace of the cube of its adjacency matrix A; the number c_4 of quadrilaterals and the number c_5 of pentagons is also expressible in terms of A.

(Harary and Manvel [HM1])

(b) The number of paths of three, four, and five points can be expressed in terms of A. (Cartwright and Gleason [CG1], and Harary and Ross [HR1])

1.19 The number of ways of arranging 2^n binary digits in a circular array so that the 2^n sequences of n consecutive digits in the arrangement are all distinct:

$$2^{2^{n-1}-n}.$$

(*Hint:* Apply Theorem (1.8.2) to the eulerian digraphs (using the term loosely because loops are present) which is obtained from the universal relation $S \times S$ on a 2-set $S = \{0, 1\}$, by taking iterated line digraphs (defined in Section 10.3).) (deBruijn [B4])

Chapter 2 | PÓLYA'S THEOREM

In order to determine the number of unlabeled graphs, the problem is reformulated so that the answer can be obtained by finding the number of orbits of the appropriate permutation group. Burnside's Lemma can then be used to express the number of orbits in terms of the number of objects fixed by permutations in the group at hand. Every permutation group has associated with it a polynomial called the "cycle index." This concept can be traced back to Frobenius as a special case of a formulation in terms of group characters. Rudvalis and Snapper [RS2] point out the connection between these generalized characters and the theorems of deBruijn [B5] and Foulkes [F1]. Redfield [R10], who discovered cycle indexes independently, devised a clever scheme (Chapter 7) which enabled him to determine the number of classes of certain matrices by forming a special product of cycle indexes. Burnside's Lemma was concealed in the proof of his enumeration theorem. Redfield's methods enabled him to count numerous interesting combinatorial structures, provided that the counting problem under consideration could be recast in the matrix form required by his theorem and provided that formulas could be derived for the relevant cycle indexes. Though admirably suited for solving certain problems, this method is somewhat difficult to apply to others because the structures to be enumerated must be interpreted as matrices. The classical enumeration theorem of Pólya, on the other hand, may be viewed as an enumerator of functions and for this reason is much easier to apply to most graphical problems. In its sweeping generality, Pólya's theorem incorporates Burnside's Lemma, and often enables one to express the complete generating function for a class of graphs in terms of an appropriate cycle index and a polynomial called the "figure counting series." Thus it is the generality, versatility, and ease with which it can be applied that make Pólya's method a most powerful tool in enumerative analysis.

2.1 GROUPS AND GRAPHS

The study of permutation groups evidently goes hand-in-hand with the study of graphs because a graph provides a "picture" of its automorphism group. Thus the group theoretic concepts required in this chapter are more easily understood in their graph-theoretic setting.

Consider a set $X = \{1, 2, \ldots, n\}$ and let A be a collection of permutations of X which is closed under multiplication. Then A is a *permutation group* with *object set* X. The *order* of A, denoted $|A|$, is the number of permutations in A and the *degree* of A is the number n of elements in the object set X. For example, consider the graph G of Figure 2.1.1, always chosen at random, whose four points consist of the set X of integers 1, 2, 3, 4. Note that the list of permutations α_i in the figure consists of all permutations of X which preserve adjacency in G. For example, points 1 and 4 are adjacent in G. The permutation (13)(2)(4) sends the points 1 and 4 to 3 and 4, and these images, 3 and 4, are also adjacent. Thus (13)(2)(4) preserves the adjacency of the points 1 and 4. Since the collection of permutations in this list is closed under multiplication, it constitutes a group. As already noticed, the

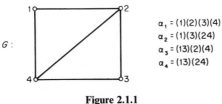

$\alpha_1 = (1)(2)(3)(4)$
$\alpha_2 = (1)(3)(24)$
$\alpha_3 = (13)(2)(4)$
$\alpha_4 = (13)(24)$

Figure 2.1.1

A graph and its group.

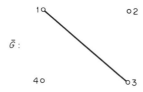

Figure 2.1.2

The complement of the random graph.

collection $\Gamma(G)$ of *all* adjacency preserving permutations of $V(G)$ is called *the group of G* or *the automorphism group of G*, and its permutations are called *automorphisms*. Thus the group of a graph is a permutation group whose objects are its points.

The *complement* $\bar{G}$ of a graph G has the same set of points as G, and two points u and v are adjacent in $\bar{G}$ if and only if they are not adjacent in G. The complement of the four point graph in Figure 2.1.1 is shown in Figure 2.1.2. The permutations which preserve adjacency in the graph of Figure 2.1.2 are the same as those for Figure 2.1.1. In fact, for any graph G, the permutations which preserve adjacency, also preserve nonadjacency and hence

$$\Gamma(\bar{G}) = \Gamma(G). \qquad (2.1.1)$$

But now we require a more subtle criterion than group isomorphism for deciding whether or not two permutation groups are the same. Consider the three labeled graphs of Figure 2.1.3, which have essentially the same groups. The only distinctions between the graphs lie in complementation and labeling. It is convenient, therefore, to identify permutation groups whose permutations are the same except for the names of the objects being permuted. Therefore we make the following definitions. Two permutation groups A, B with object sets X and Y respectively are *isomorphic*, written $A \cong B$, if there is a function h from A onto B such that for all α_1, α_2 in A

$$h(\alpha_1\alpha_2) = h(\alpha_1)h(\alpha_2), \qquad (2.1.2)$$

If there is also a 1–1 map φ from X onto Y such that for each α in A and each x in X

$$\varphi(\alpha x) = h(\alpha)\varphi(x), \qquad (2.1.3)$$

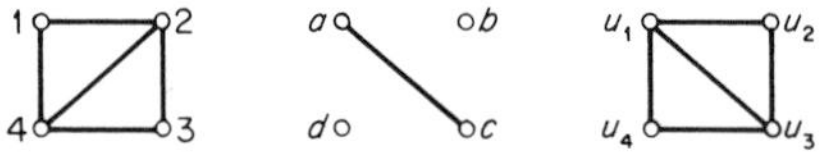

Figure 2.1.3

Three graphs with identical groups.

then A and B are *identical* and we write $A = B$. Thus, the map φ simply changes the labels, or names, of the objects of A to those of B. It is now easily seen that the groups of all three graphs in Figure 2.1.3 are identical.

2.2 THE CYCLE INDEX OF A PERMUTATION GROUP

Let A be a permutation group with object set $X = \{1, 2, \ldots, n\}$. It is well known that each permutation α in A can be written uniquely as a product of disjoint cycles and so for each integer k from 1 to n we let $j_k(\alpha)$ be the number of cycles of length k in the disjoint cycle decomposition of α. Then the *cycle index* of A. denoted $Z(A)$, (Z for the word *Zyklenzeiger* used by Pólya [P8]) is the polynomial in the variables $s_1, s_2, \ldots, s_n$ defined by

$$Z(A) = |A|^{-1} \sum_{\alpha \in A} \prod_{k=1}^{n} s_k^{j_k(\alpha)}. \qquad (2.2.1)$$

When it is necessary to display the variables, we shall write $Z(A; s_1, s_2, \ldots, s_n)$ instead of $Z(A)$. Redfield [R10] called this polynomial a "group-reduction function" and Pólya [P8], who discovered the concept independently, named it the "cycle index."

To provide an example, we consider the symmetric group S_n on n objects. For $n = 3$, we observe that the identity permutation $(1)(2)(3)$ has three cycles of length 1, resulting in the term s_1^3. The three permutations $(1)(23)$, $(2)(13)$, and $(3)(12)$ each have one cycle of length 1 and one of length 2, and so one term is obtained, $3s_1 s_2$. Finally, the two permutations (123) and (132) contribute $2s_3$. Thus we have

$$Z(S_3) = (1/3!)(s_1^3 + 3s_1 s_2 + 2s_3). \qquad (2.2.2)$$

Throughout the rest of this book we shall make very frequent use of the explicit formulas which follow for the cycle indexes of the five most famous permutation groups: symmetric, alternating, cyclic, dihedral, and identity. Both Redfield [R10] and Pólya [P8] expressed $Z(S_n)$ in terms of the partitions of n. Note that each permutation α of n objects can be associated with the partition of n which has, for each k from 1 to n, exactly $j_k(\alpha)$ parts equal to k. We shall denote a partition of n by the vector $(j) = (j_1, j_2, \ldots, j_n)$ where j_k is the number of parts equal to k. Thus

$$n = \sum_{k=1}^{n} k j_k. \qquad (2.2.3)$$

Let $h(j)$ be the number of permutations in S_n whose cycle decomposition determines the partition (j), so that for each k, $j_k = j_k(\alpha)$. Then it is easy to see that

$$h(j) = n!/\prod k^{j_k} j_k !. \tag{2.2.4}$$

Thus the cycle index $Z(S_n)$ takes the form shown in the next theorem.

Theorem   The cycle index of the symmetric group is given by

$$Z(S_n) = (1/n!) \sum_{(j)} h(j) \prod_{k=1}^{n} s_k^{j_k}, \tag{2.2.5}$$

where the sum is over all partitions (j) of n, and $h(j)$ is given by (2.2.4).

The following corollary gives the cycle index of the *alternating group* A_n which consists of all the even permutations in S_n.

Corollary   The cycle index of the alternating group is given by

$$Z(A_n) = Z(S_n) + Z(S_n; s_1, -s_2, s_3, -s_4, \ldots). \tag{2.2.6}$$

To illustrate, note that from (2.2.2) we have

$$Z(S_3; s_1, -s_2, s_3) = (1/3!)(s_1^3 - 3s_1 s_2 + 2s_3), \tag{2.2.7}$$

and on adding (2.2.2) and (2.2.7) we obtain

$$Z(A_3) = \tfrac{1}{3}(s_1^3 + 2s_3). \tag{2.2.8}$$

It is often convenient to express $Z(S_n)$ in terms of $Z(S_k)$ with $k < n$. For this purpose we define $Z(S_0) = 1$, and the recursive formula, whose inductive proof is straightforward, can then be stated as follows.

Theorem   The cycle index of the symmetric group satisfies the recurrence relation

$$Z(S_n) = n^{-1} \sum_{k=1}^{n} s_k Z(S_{n-k}). \tag{2.2.9}$$

The cyclic group of degree n, denoted C_n, is generated by the cycle $(123 \cdots n)$. Redfield provided the following formula for $Z(C_n)$ using the Euler φ-function.

Theorem   The cycle index of the cyclic group C_n is given by

$$Z(C_n) = n^{-1} \sum_{k|n} \varphi(k) s_k^{n/k}. \tag{2.2.10}$$

The *dihedral group of degree n*, denoted D_n, is generated by the cycle $(123 \cdots n)$ and the reflection $(1\ n)(2\ n-1)(3\ n-2)\cdots$. Its cycle index can be expressed in terms of $Z(C_n)$.

Corollary The cycle index of the dihedral group D_n is given by

$$Z(D_n) = \tfrac{1}{2}Z(C_n) + \begin{cases} \frac{1}{2}s_1 s_2^{(n-1)/2} & n \ \text{ odd} \\[2mm] \frac{1}{4}(s_2^{n/2} + s_1^2 s_2^{(n-2)/2}) & n \ \text{ even.} \end{cases} \tag{2.2.11}$$

To illustrate, one may use (2.2.10) and (2.2.11) to obtain $Z(C_3)$ and $Z(D_3)$, but note that $C_3 = A_3$ and $D_3 = S_3$. Therefore $Z(C_3)$ and $Z(D_3)$ are also given by (2.2.8) and (2.2.2) respectively. Our group notation is completed by letting E_n denote the *identity group* on n objects, so that

$$Z(E_n) = s_1^n. \tag{2.2.12}$$

We observe next that the cycle index does not determine a unique permutation group. That is, two permutation groups A and B need not be identical to share the same cycle index. In fact, they may even be nonisomorphic and yet have the same cycle index and as the following rather accurate translation from Pólya [P8, p. 176] demonstrates:

> It is of interest to remark that two combinatorially equivalent permutation groups (having the same cycle index) need not be identical. In fact they need not even be isomorphic. Namely, let p be an odd prime and $m \geq 3$ be an integer ($p = m = 3$ is the simplest example). It is well known (see Burnside [B7, p. 143]) that there is a nonabelian group of order p^m in which every element except the identity has order p. Let B be the regular representation of this group as a permutation group. Let A be the regular representation of the abelian group of order p^m and type $(p, p, \ldots, p)$. Then A and B are permutation groups of order and degree $p^m = d$ with the same cyclic index
>
> $$d^{-1}(s_1^d + (d-1)s_p^{d/p})$$
>
> for each permutation of A and B other than the identity contains p^{m-1} cycles of length p.

This section is concluded with a binary operation on permutation groups together with the relevant cycle index formula. Let A and B be groups with disjoint object sets X and Y respectively. The *product*† of A and B, denoted AB, is a permutation group with object set $X \cup Y$. Each pair of permutations, α in A and β in B, determines a permutation, denoted $\alpha\beta$, in AB such that for each z in $X \cup Y$,

$$\alpha\beta(z) = \begin{cases} \alpha z, & z \in X \\ \beta z, & z \in Y. \end{cases} \tag{2.2.13}$$

† We formerly [H1] called this product the "sum" and denoted it by $A + B$.

Thus AB has degree $|X| + |Y|$ and order $|A||B|$. We denote the product $AA \cdots A$ of m copies of the group by A^m. Pólya [P8] observed the elementary but useful fact that the cycle index of a product is the product of the cycle indexes of the constituent groups.

Theorem   The cycle index of the product AB is given by

$$Z(AB) = Z(A)Z(B). \qquad (2.2.14)$$

Of course the complete graph K_n on n points has S_n as its group. Furthermore the graph G whose only connected components are K_n and K_m, with $m \neq n$, has group $\Gamma(G) = S_n S_m$. Hence from (2.2.14), $Z(\Gamma(G)) = Z(S_n)Z(S_m)$. Since it will be necessary to refer to the cycle index of the group of a graph, we often simplify the notation by writing $Z(G)$ instead of $Z(\Gamma(G))$. For example, $Z(K_n)$ is given by (2.2.4) and (2.2.5).

2.3   BURNSIDE'S LEMMA

The three lemmas discussed next form the basis for numerous solutions to counting problems for unlabeled graphs. Though apparently known to Frobenius, Schur, and others we refer to them as the lemmas of Burnside [B7]. Let A be a permutation group with object set $X = \{1, 2, \ldots, n\}$. Then x and y in X are called *A-equivalent* or *similar* if there is a permutation α in A such that $\alpha x = y$. It is a classical and immediate result that this is an equivalence relation and the equivalence classes are called the *orbits* or *transitivity systems* of A.

For each x in X, let

$$A(x) = \{\alpha \in A | \alpha x = x\}. \qquad (2.3.1)$$

Thus $A(x)$ is called the *stabilizer* of x. Note that whenever x and y belong to the same orbit, $A(x)$ and $A(y)$ are conjugate subgroups of A, and hence $|A(x)| = |A(y)|$. We now show that for any element y of an orbit Y of A,

$$|A| = |A(y)||Y|, \qquad (2.3.2)$$

that is, the number of elements in the orbit of y is the index of the stabilizer of y in A. To see this, we first express A as a union of right cosets modulo $A(y)$:

$$A = \bigcup_{i=1}^{m} \alpha_i A(y).$$

It only remains to observe the natural 1–1 correspondence between these cosets and the elements of Y. For each $i = 1$ to m, we associate the coset $\alpha_i A(y)$ with the element $\alpha_i(y)$ in Y. For $i \neq j$, we have $\alpha_i(y) \neq \alpha_j(y)$, because otherwise $\alpha_j^{-1}\alpha_i$ is an element of $A(y)$ and hence α_i is an element of $\alpha_j A(y)$, thus contradicting the fact that $\alpha_i A(y) \cap \alpha_j A(y) = \emptyset$. Therefore this correspondence is 1–1. For any object y' in Y we have $\alpha(y) = y'$ for some permutation α in A. From the coset decomposition of A, it follows that $\alpha = \alpha_i \gamma$ with γ in $A(y)$. Hence $y' = \alpha_i(y)$ and thus every element of Y corresponds to some coset. Therefore m is the number of elements in Y and (2.3.2) is proved.

Now we are prepared for the first lemma which provides a formula for the number $N(A)$ of orbits of A in terms of the average number of fixed points of the permutations in A.

Burnside's Lemma   The number $N(A)$ of orbits of A is given by

$$N(A) = |A|^{-1} \sum_{\alpha \in A} j_1(\alpha). \tag{2.3.3}$$

Proof   Let $X_1, X_2, \ldots, X_m$ be the orbits of A and for each $i = 1$ to m, let x_i be an element of the ith orbit, X_i. Then from (2.3.2) we have

$$N(A)|A| = \sum_{i=1}^{m} |A(x_i)|\,|X_i|. \tag{2.3.4}$$

We have seen that if x and x_i are in the same orbit, then $|A(x)| = |A(x_i)|$. Hence the right side of (2.3.4) can be altered to obtain

$$N(A)|A| = \sum_{x \in X} |A(x)|, \tag{2.3.5}$$

or in other notation

$$N(A)|A| = \sum_{x \in X} \sum_{\alpha \in A(x)} 1. \tag{2.3.6}$$

Now on interchanging the order of summation on the right side of (2.3.6) and modifying the summation indices accordingly, we have

$$N(A)|A| = \sum_{\alpha \in A} \sum_{x = \alpha x} 1, \tag{2.3.7}$$

but $\sum_{x = \alpha x} 1$ is just $j_1(\alpha)$. Thus the proof is completed on division by $|A|$.   //

To illustrate, consider the graph G in Figure 2.3.1. Using the product notation, the group of G may be expressed as $\Gamma(G) = S_1^3 S_2^2$. Now $\Gamma(G)$ has order 4 and each permutation fixes the three points 3, 5 and 7. Let the permutations be denoted by

$$\alpha_1 = (1)(2)(3)(4)(5)(7) \qquad \alpha_3 = (46)(1)(2)(3)(5)(7)$$
$$\alpha_2 = (12)(3)(4)(5)(6)(7) \qquad \alpha_4 = (12)(46)(3)(5)(7). \tag{2.3.8}$$

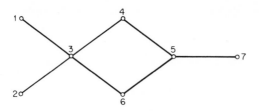

Figure 2.3.1

A graph with three fixed points.

Then $j_1(\alpha_1) = 7$, $j_1(\alpha_2) = j_1(\alpha_3) = 5$ and $j_1(\alpha_4) = 3$. Thus $N(\Gamma(G)) = \frac{1}{4}(7 + 5 + 5 + 3) = 5$. It is clear of course that the orbits are $\{3\}, \{5\}, \{7\}, \{1, 2\}$, and $\{4, 6\}$. Note that the number of orbits is precisely the number of ways in which G can be rooted. To obtain all rootings of G, one simply chooses one point from each of the orbits as a root.

Occasionally we will need to restrict A to a subset Y of X, where Y is a union of orbits of A. Therefore we denote by $A|Y$ the set of permutations on Y obtained by restricting those of A to Y. For each α in A, the number of elements in Y fixed by α is denoted by $j_1(\alpha|Y)$. Then we can state the following consequence of (2.3.3).

Restricted Form of Burnside's Lemma

$$N(A|Y) = |A|^{-1} \sum_{\alpha \in A} j_1(\alpha|Y). \qquad (2.3.9)$$

Next we provide a slight generalization of (2.3.3) called the Weighted Form of Burnside's Lemma. Let R be any commutative ring containing the rationals and let w be a function, called the *weight function*, from the object set X of A into the ring R. In practice the weight function is constant on the orbits of A. Hence in this case we can define the weight of any orbit X_i to be the weight of any element in the orbit. For each orbit X_i, we denote the weight of X_i by $w(X_i)$, and by definition, $w(X_i) = w(x)$ for any x in X_i.

Weighted Form of Burnside's Lemma   The sum of the weights of the orbits of A is given by

$$\sum_{i=1}^{m} w(X_i) = |A|^{-1} \sum_{\alpha \in A} \sum_{x = \alpha x} w(x). \qquad (2.3.10)$$

The proof is similar to the proof of (2.3.3) and is omitted.

We consider again the graph G in Figure 2.3.1 to illustrate this lemma, and to display a sum of cycle indexes in a way which will be used effectively in Chapter 8. For each point k of G, we define the weight $w(k)$ to be the cycle index of the stabilizer of k in $\Gamma(G)$. Thus

$$w(1) = \tfrac{1}{2}(s_1^7 + s_1^5 s_2) \tag{2.3.11}$$

and

$$w(3) = \tfrac{1}{4}(s_1^7 + 2s_1^5 s_2 + s_2^2 s_3). \tag{2.3.12}$$

Note that

$$w(1) = w(2) = w(4) = w(6) \quad \text{and} \quad w(3) = w(5) = w(7); \tag{2.3.13}$$

thus in particular w is constant on the orbits.

We sketch the verification of (2.3.10) for this example by observing that the sum of the orbit weights is $w(1) + w(3) + w(4) + w(5) + w(7) = 2w(1) + 3w(3)$. Now the right side of (2.3.10) is the sum

$$\tfrac{1}{4} \sum_{i=1}^{4} \sum_{x = \alpha_i x} w(x) = \tfrac{1}{4} \left\{ \sum_{x = \alpha_1 x} w(x) + \cdots + \sum_{x = \alpha_4 x} w(x) \right\}$$

$$= \tfrac{1}{4} \left\{ \left[\sum_{k=1}^{7} w(k) \right] + \left[\sum_{k=3}^{7} w(k) \right] + \cdots \right\},$$

where the last two terms in the above equation are found at once by inspecting α_3 and α_4 in (2.3.8). Using (2.3.13), this sum may be rewritten as $2w(1) + 3w(3)$, which we saw is the sum of the orbit weights, and the verification is finished.

Similarly, the cycle index sum for all the different rooted graphs obtained from any graph G can be obtained in terms of the weights of the fixed points of $\Gamma(G)$.

2.4 POLYA'S THEOREM

Since most applications of Pólya's main enumeration theorem have required only the one-variable version and since the theorem is also more easily understood in this case, we shall not provide details of the usual generalization to n variables.

First we introduce the power group [HP4], which we will encounter again and again later in this book. Let A be a permutation group with object set $X = \{1, 2, \ldots, n\}$ and let B be a finite permutation group with a countable object set Y of at least two elements. Then the *power group* denoted B^A has the collection Y^X of functions from X into Y as its object set. The permutations of B^A consist of all ordered pairs, written $(\alpha; \beta)$, of permutations α

in A and β in B. The image of any function f in Y^X under $(\alpha; \beta)$ is given by

$$((\alpha; \beta)f)(x) = \beta f(\alpha x), \tag{2.4.1}$$

for each x in X.

In order to lead to the classical Pólya enumeration formula, we shall take $B = E$, the identity group on Y. Now consider the power group E^A acting on Y^X. Let $w: Y \rightarrow \{0, 1, 2, \ldots\}$ be a function whose range is the set of nonnegative integers, and for which $|w^{-1}(k)| < \infty$ for all k. In particular, for each $k = 0, 1, 2, \ldots$ let

$$c_k = |w^{-1}(k)| \tag{2.4.2}$$

be the number of "figures" with weight k.

Then the elements y in Y which have $w(y) = k$, are said to have weight k and w is called a *weight function*. Further the series in the indeterminate x,

$$c(x) = \sum_{k=0}^{\infty} c_k x^k, \tag{2.4.3}$$

which enumerates the elements of Y by weight, is called the "figure counting series."

The *weight of a function* f in Y^X is defined by

$$w(f) = \sum_{x \in X} w(f(x)), \tag{2.4.4}$$

and it is then easily seen that functions in the same orbit of the power group E^A have the same weight. Therefore the *weight* $w(F)$ *of an orbit* F of E^A is the weight of any f in F. Since $|w^{-1}(k)| < \infty$ for each $k = 0, 1, 2, \ldots$, there are only a finite number of orbits of each weight. Hence we let C_k be the number of orbits of weight k. Then the series in the indeterminate x,

$$C(x) = \sum_{k=0}^{\infty} C_k x^k, \tag{2.4.5}$$

is called the "function counting series," or the "configuration counting series" following Pólya [P8]. Now we can finally state the main theorem of this book which expresses $C(x)$ in terms of $Z(A)$ and $c(x)$. In this formula $Z(A, c(x))$ is an abbreviation for $Z(A; c(x), c(x^2), c(x^3), \ldots)$.

Theorem (Pólya's Enumeration Theorem) The function counting series $C(x)$ is determined by substituting for each variable s_k in $Z(A)$, the figure counting series $c(x^k)$. Symbolically

$$C(x) = Z(A, c(x)). \tag{2.4.6}$$

This result is used so often throughout graphical enumeration that we will frequently write PET for brevity instead of Pólya's Enumeration Theorem.

Proof Let ε be the identity permutation on Y. Then for each α in A, and each $k = 0, 1, 2, \ldots$, we can let $\varphi(\alpha, k)$ be the number of functions of weight k fixed by $(\alpha; \varepsilon)$. Now for each k, on restricting the power group E^A to the functions of weight k and applying the Restricted Form (2.3.9) of Burnside's Lemma, we have

$$C_k = |A|^{-1} \sum_{\alpha \in A} \varphi(\alpha, k). \tag{2.4.7}$$

Therefore

$$C(x) = \sum_{k=0}^{\infty} |A|^{-1} \sum_{\alpha \in A} \varphi(\alpha, k) x^k \tag{2.4.8}$$

and on interchanging the order of summation, we have

$$C(x) = |A|^{-1} \sum_{\alpha \in A} \sum_{k=0}^{\infty} \varphi(\alpha, k) x^k. \tag{2.4.9}$$

Now $\sum_{k=0}^{\infty} \varphi(\alpha, k) x^k$ is the counting series for all functions fixed by $(\alpha; \varepsilon)$ and we seek an alternative form for this series.

Suppose f in Y^X is fixed by $(\alpha; \varepsilon)$. Then $(\alpha; \varepsilon) f(x) = f(x)$ for all x in X, but from (2.4.1) we have $(\alpha; \varepsilon) f(x) = \varepsilon f(\alpha x)$. Thus we must have $f(\alpha x) = f(x)$ for all x, and hence f must be constant on the disjoint cycles of α. Conversely, all functions constant on the cycles of α are fixed by $(\alpha; \varepsilon)$.

Let z_r be a cycle of length r in α. If f sends the elements of z_r to one of the c_k elements of Y of weight k, then the contribution to the weight of f is rk. Then it can be seen that the series

$$c(x^r) = \sum_{k=0}^{\infty} c_k x^{rk} \tag{2.4.10}$$

has as the coefficient of x^{rk} for each k, the number of ways f can be defined on the elements of z_r so that f is fixed by $(\alpha; \varepsilon)$ and the contribution to $w(f)$ is rk. It follows that $c(x^r)^{j_r(\alpha)}$ enumerates by weight the ways of defining fixed functions on all the cycles of length r in α.

On considering all cycles of α, we can then express the series for fixed functions as the product

$$\sum_{k=0}^{\infty} \varphi(\alpha, k) x^k = \prod_{k=1}^{n} c(x^k)^{j_k(\alpha)}. \tag{2.4.11}$$

Now (2.4.6) follows from (2.4.9), (2.4.11), and the definition of $Z(A)$, and the proof is completed.   //

Figure 2.4.1

Two labeled necklaces.

To illustrate the theorem we shall consider a "necklace problem." In Figure 2.4.1 we have shown two necklaces, each with four beads. Each of these has its beads labeled with the elements of $X = \{1, 2, 3, 4\}$, and each has two white beads and two black. Clearly the number of labeled necklaces constructed from black and white beads is 2^4. To obtain the number of unlabeled necklaces, we must identify necklaces such as those in Figure 2.4.1 when one necklace differs from another only by a reflection or rotation of the string of beads. If we let $Y = \{a, b\}$, then each function f from X to Y corresponds to a labeled necklace in which bead number k in X has "color" $f(k)$. Thus the necklace represented by f has $|f^{-1}(a)|$ beads of one color and $|f^{-1}(b)|$ beads of the other color. Now let the identity group E_2 act on Y. To remove the labels, two necklaces should be identified whenever their corresponding functions are in the same orbit of the power group $E_2^{D_4}$. If we define $w(a) = 0$ and $w(b) = 1$, then $1 + x$ is the counting series for Y and a function of weight k represents a necklace with $4 - k$ white beads and k black beads. Therefore the function counting series $C(x)$ here enumerates unlabeled necklaces and the coefficient of x^k is the number of such with k black beads. From (2.4.6) in the PET, then,

$$C(x) = Z(D_4, 1 + x). \tag{2.4.12}$$

From (2.2.11) we have

$$Z(D_4) = \tfrac{1}{8}(s_1^4 + 2s_1^2 s_2 + 3s_2^2 + 2s_4). \tag{2.4.13}$$

Then on carrying out the details of the substitution of the figure counting series $1 + x$ in $Z(D_4)$ we have

$$C(x) = 1 + x + 2x^2 + x^3 + x^4. \tag{2.4.14}$$

The six unlabeled necklaces with four beads of two colors are now shown in Figure 2.4.2.

Figure 2.4.2

The necklaces with four beads.

The total number of necklaces is, of course, $C(1)$ and hence the total number may be obtained by evaluating the figure counting series $1 + x$ at $x = 1$ and substituting 2 for each variable s_k in $Z(D_4)$. In general, whenever the figure counting series $c(x)$ is a polynomial, so is the function counting series $C(x)$. Then $C(1)$, the number of orbits of functions without regard to weight is obtained on substituting $c(1)$ for each variable in the cycle index at hand.

Corollary The number of orbits of functions determined by the power group E_m^A is obtained by substituting the integer m for each variable in $Z(A)$:

$$N(E_m^A) = Z(A, m). \qquad (2.4.15)$$

We shall provide only the statement of the PET for n variables. The proof follows the same course as above for the one-variable case. Let N be the set of nonnegative integers and let $N^n = N \times \cdots \times N$ be the cartesian product of n copies of N. As before, the power group E^A has object set Y^X, and $w: Y \to N^n$, the weight function, has the property that for each z in N^n, $|w^{-1}(z)| < \infty$. With component-wise addition in N^n, weights of functions in Y^X and orbits of E^A are defined as before. By definition of the figure counting series, $c(x_1, x_2, \ldots, x_n)$ has $|w^{-1}(r_1, r_2, \ldots, r_n)|$ as the coefficient of $x_1^{r_1} x_2^{r_2} \cdots x_n^{r_n}$, and the function counting series $C(x_1, x_2, \ldots, x_n)$ has as the coefficient of $x_1^{t_1} x_2^{t_2} \cdots x_n^{t_n}$, the number of orbits of weight $(t_1, t_2, \ldots, t_n)$. We denote by $Z(A, c(x_1, x_2, \ldots, x_n))$ the polynomial obtained with each variable s_k in $Z(A)$ is replaced by $c(x_1^k, x_2^k, \ldots, x_n^k)$. Then the more general result giving the PET in n variables can be expressed as follows.

Theorem If $c(x_1, x_2, \ldots, x_n)$ is the figure counting series for Y, then the orbits of functions in Y^X determined by the power group E^A are enumerated by weight with $C(x_1, x_2, \ldots, x_n)$ and

$$C(x_1, x_2, \ldots, x_n) = Z(A, c(x_1, x_2, \ldots, x_n)). \qquad (2.4.16)$$

To illustrate this theorem, we return to the necklace problem. This time we wish to determine the enumerator of necklaces with four beads when three colors of beads are available. Therefore we let $Y = \{a, b, c\}$ and we can consider any function f from X to Y as representing a necklace with $|f^{-1}(a)|$ red beads, $|f^{-1}(b)|$ white beads, and $|f^{-1}(c)|$ blue beads. If we let $w(a) = (0, 0)$, $w(b) = (1, 0)$, and $w(c) = (0, 1)$, then

$$w(f) = \sum_{x \in X} w(f(x)), \qquad (2.4.17)$$

and $w(f)$ is an ordered pair whose first coordinate is the number of white beads in the necklace and the second is the number of blue. The number of red beads is, of course, just $|X|$ minus the number of white and blue. Now, by definition, the figure counting series is $c(x) = 1 + x_1 + x_2$. Hence by the theorem, the necklace enumerator is

$$C(x_1, x_2) = Z(D_4, 1 + x_1 + x_2). \tag{2.4.18}$$

On carrying out the details of the substitution we obtain

$$C(x_1, x_2) = 1 + x_1 + 2x_1^2 + x_1^3 + x_1^4 + x_2 + 2x_2^2 + x_2^3 + x_2^4$$
$$+ 2x_1 x_2 + 2x_1^2 x_2 + x_1^3 x_2 + 2x_1 x_2^2 + x_1 x_2^3 + 2x_1^2 x_2^2. \tag{2.4.19}$$

As a check, one can easily compute the coefficient sum for $C(x_1, x_2)$ by evaluating $Z(D_4, 3)$, which is 21; then compare (2.4.19).

2.5 THE SPECIAL FIGURE SERIES $1 + x$

There is a rather natural corollary to the PET which stipulates the significance of the coefficients of the polynomial obtained when $1 + x$ is substituted into the cycle index of an arbitrary permutation group A. This observation, although quite easy to demonstrate, is exceptionally powerful because every enumeration formula obtained from the PET by using the figure series $1 + x$ is necessarily a special case thereof. These results will include the counting of necklaces, graphs, digraphs, rooted graphs, and bicolored graphs. Thus we defer such applications of the corollary until later.

We note now its impact on the five special groups whose cycle indexes are given in equations (2.2.5, 6, 10, 11, 12). Just as for individual elements of X two r-sets $S = \{x_1, \ldots, x_r\}$ and $S' = \{x'_1, \ldots, x'_r\}$ in X are called A-equivalent if for some $\alpha \in A$, $\alpha S = S'$.

(2.5.1) **Corollary to PET** The coefficient of x^r in $Z(A, 1 + x)$ is the number of A-equivalence classes of r-sets of X.

Proof In the figure counting series $1 + x$, the term $1 = x^0$ can indicate the absence of an object in X while $x = x^1$ stands for its presence. Thus x^r means that r distinct objects, forming an r-set, are present. The corollary now follows at once from the PET.   //

In view of this corollary, we see that a permutation group A is transitive if and only if the coefficient of x in $Z(A, 1 + x)$ is 1. Furthermore this polynomial always has end-symmetry of its coefficients because the number of classes of r-sets and $(n - r)$-sets are equal. For the symmetric group, it follows from the definition of S_n that there exists a permutation taking a given r-set onto any other. For A_n, one only need note that an even permutation can be found which effects this mapping. The next two formulas can also be verified by the brute force substitution of $1 + x$ into the respective cycle indexes:

$$Z(S_n, 1 + x) = 1 + x + x^2 + \cdots + x^n \qquad (2.5.2)$$

$$Z(A_n, 1 + x) = 1 + x + x^2 + \cdots + x^n. \qquad (2.5.3)$$

The identity group of course produces the binomial coefficients:

$$Z(E_n, 1 + x) = \sum_{r=0}^{n} \binom{n}{r} x^r. \qquad (2.5.4)$$

The cyclic and dihedral groups are a bit more complicated. It is a routine matter to make a formal substitution of $1 + x^k$ for s_k in (2.2.10) to obtain

$$Z(C_n, 1 + x) = n^{-1} \sum_{k|n} \varphi(k)(1 + x^k)^{n/k}. \qquad (2.5.5)$$

However a similar substitution into (2.2.11) results in a less elegant-looking equation. Rather than write this formula mechanically, we note that we have already encountered it for $n = 4$ in (2.4.14) where the polynomial so obtained counts four-bead necklaces with a specified number of beads of each of two colors. For arbitrary n, the result then gives the number of types of two-color, n-bead necklaces.

2.6   ONE–ONE FUNCTIONS

It is now logically convenient to have at our disposal a theorem of Pólya which expresses the number of weighted 1–1 functions in terms of the cycle indexes of the symmetric and alternating groups and a figure counting series. We shall use this result later to relate the generating function for trees to that for rooted trees.

Let $c(x)$ be the series that enumerates the elements of any set Y according to weight and let the identity group E have object set Y. Now consider the power group E^A restricted to the 1–1 functions in Y^X. If $C(x)$ is the counting series for the orbits of 1–1 functions determined by E^A, we seek to express $C(x)$ in terms of $c(x)$. We shall do this first with $A = S_n$; then the solution to the general problem follows quickly. Note that the orbits of 1–1 functions determined by E^{S_n} correspond to n-combinations or n-subsets of the elements of Y. Following Pólya we have used $Z(A_n - S_n)$ as an abbreviation for $Z(A_n) - Z(S_n)$, and we set $Z(A_0 - S_0) = 1$.

Theorem   The generating function $C(x)$ which enumerates 1–1 functions from n indistinguishable elements into a collection of objects with figure counting series $c(x)$ is given by

$$C(x) = Z(A_n - S_n, c(x)). \qquad (2.6.1)$$

Before proving the theorem we shall illustrate its use with an example for $n = 3$. Let $c(x)$ be the generating function for the set Y of connected graphs, so that the coefficient of x^p in $c(x)$ is the number of connected graphs of order p. It is known that the first few terms of $c(x)$ are

$$c(x) = x + x^2 + 2x^3 + 6x^4 + 21x^5 + 112x^6 + \cdots. \qquad (2.6.2)$$

Next let $C(x)$ be the generating function for graphs that have exactly three components, all different. Consider the power group E^{S_3} with object set Y^X. Then the orbits of 1–1 functions determined by E^{S_3} correspond precisely to the graphs enumerated by $C(x)$. Furthermore, the weight of each orbit is the order of the graph to which the orbit corresponds.

The cycle index formula for $Z(A_3 - S_3)$ is already available in (2.2.7). Therefore on substituting $c(x^k)$ for each variable s_k in this formula, the first few terms of $C(x)$ are seen to be

$$C(x) = 2x^6 + 7x^7 + 34x^8 + \cdots. \qquad (2.6.3)$$

It is left to the reader to verify some of these coefficients by drawing the corresponding graphs.

Proof   To prove (2.6.1), recall that $c(x)$ enumerates the elements of any set Y by weight and E^{S_n} has object set Y^X. It follows from PET that the counting series for orbits of *all* functions determined by E^{S_n} is simply $Z(S_n, c(x))$. Therefore it is sufficient to show that $Z(A_n, c(x))$ counts these orbits of 1–1 functions twice and all others just once.

We first note that the number of orbits of 1–1 functions from X to itself determined by $E_n^{A_n}$ is 2. This is an immediate consequence of the easily demonstrated fact that any two such 1–1 functions are in the same orbit of $E_n^{A_n}$ if and only if they are both odd or both even. Thus $Z(A_n, c(x))$ counts twice those orbits of E^{S_n} which consist of 1–1 functions.

Next we show that orbits of E^{S_n} which consist of functions not 1–1 are counted only once. To do this, consider such an orbit and any two functions f and g in it. Then there is a permutation α in S_n such that for all x in X, $f(x) = g(\alpha x)$. We need to show that f and g are in the same orbit of E^{A_n}.

This follows from the equation $f(x) = g(\alpha x)$ if α is even. Suppose, on the other hand, that α is odd. Since f is not 1–1, for some x_1, x_2 in X with $x_1 \neq x_2$ we have $f(x_1) = f(x_2)$. Let β be the permutation that interchanges x_1 and x_2 and fixes all other elements of X. Of course β, being a transposition is odd. Thus $\alpha\beta$ is even, and for all x in X, $f(x) = g(\alpha\beta x)$. Thus f and g are also in the same orbit of E^{A_n} and hence $Z(A_n, c(x))$ counts orbits of E^{S_n} which consist of functions not 1–1 exactly once. $/\!/$

The following corollary treats the general case, in which the orbits of 1–1 functions on n objects are determined not by the symmetric group S_n but by any group A of degree n.

Corollary The generating function $C(x)$ that enumerates 1–1 functions determined by the figure counting series $c(x)$ and any permutation group A of degree n is given by

$$C(x) = \frac{n!}{|A|} Z(A_n - S_n, c(x)). \tag{2.6.4}$$

Proof Recall that $c(x)$ enumerates the elements of Y by weight and $Z(A_n - S_n, c(x))$ counts by weight the subsets of Y that consist of n elements. As usual E is the identity group with object set Y and A is a permutation group of degree n with object set X. Consider any n-subset Y_1 of Y. Then we seek to establish that the number of orbits of E_n^A restricted to the 1–1 functions in Y_1^X is $n!/|A|$. But this conclusion follows immediately from the Restricted Form of Burnside's Lemma because the only permutation in E_n^A that fixes any 1–1 function in Y_1^X is the identity permutation, which fixes all $n!$ of them. $/\!/$

In our applications of this theorem, it is often necessary to sum the polynomials $Z(A_n - S_n)$. We sometimes write $Z(A_\infty - S_\infty)$ instead of $\sum_{n=0}^{\infty} Z(A_n - S_n)$. Riordan [R15] established the following formula, which is analogous to (3.1.1),

$$Z(A_\infty - S_\infty, f(x)) = \exp\left\{ \sum_{k=1}^{\infty} (-1)^{k+1} f(x^k)/k \right\}. \tag{2.6.5}$$

EXERCISES

2.1 Prove that Burnside's Lemma (2.3.3) gives the number of orbits determined by any group represented as a permutation group. Specifically, let A be any (abstract) group and suppose $\varphi: A \to B$ is a homomorphism from A onto the permutation group B. Then the number $N(B)$ of orbits of B is given by

$$N(B) = |A|^{-1} \sum_{\alpha \in A} j_1(\varphi(\alpha)).$$

2.2 How many orbits of functions are determined by the power group $E_n^{S_m}$? How many are determined by E_n^A with $A = A_m, C_m$, or D_m?

2.3 Find two nonisomorphic permutation groups A and B of smallest order such that $Z(A) = Z(B)$.   (Pólya [P8])

2.4 Prove that the set of permutations in the power group B^A is closed under multiplication.

2.5 How many necklaces are there with five beads when three colors are available?

2.6 If the vertices of a cube are colored using three different colors, how many different cubes are possible?

2.7 If the faces of a cube are colored using four different colors, how many different cubes result?

2.8 If five different colors are available and each line of the graph in Figure 2.3.1 assumes one of these colors, how many different colored graphs are possible?

2.9 If G is any graph, how may the coefficient of $Z(\Gamma(G), 1 + x)$ be interpreted? Illustrate using the graph in Figure 2.3.1.

2.10 When the object set of the power group B^A is restricted to 1–1 functions, the group is denoted by B^{A*}. Find a formula for the cycle index $Z(B^{A*})$.   (Harary and Palmer [HP5])

2.11 Find two isomorphic but not identical permutation groups of degree 6 and order 4 which share the same cycle index.   (Redfield [R10])

2.12 The cycle index of the alternating group A_n given in (2.2.6) can be expressed in the form:

$$Z(A_n) = \sum_{(j)} \frac{[1 + (-1)^{j_2 + j_4 + \cdots}]}{\prod k^{j_k} j_k!} \prod s_k^{j_k}.$$

2.13 $\sum_{n=0}^{\infty} Z(S_n) = \exp[\sum_{k=1}^{\infty} (s_k/k)].$

Chapter 3 | TREES

We have seen in Chapter 1 several methods for determining the number p^{p-2} of labeled trees of order p. We shall now consider the more difficult problem of finding the number of *unlabeled* trees, that is, the number of isomorphism classes of trees with a given number of points. The techniques that we shall use can be readily adapted not only for finding generating functions for trees with various specified properties but also for counting treelike structures.

For example, in this chapter we shall find generating functions for rooted trees, trees, forests, oriented trees, homeomorphically irreducible trees, identity trees, unicyclic graphs, functional digraphs, cacti, and 2-trees. The first few coefficients of many of these generating functions may also be found in Sloane's book of sequences [S4].

3.1 ROOTED TREES

It seems impossible to enumerate trees without first enumerating rooted trees. Therefore we begin by using the Pólya method of the preceding chapter

51

to obtain the generating function for rooted trees. We then employ a "dissimilarity characteristic theorem" due to Otter [O4] to relate this function to the series that counts trees.

We will make use of the following well-known identity which sums the cycle indexes of all the symmetric groups:

$$\sum_{n=0}^{\infty} Z(S_n, f(x)) = \exp \sum_{k=1}^{\infty} f(x^k)/k. \tag{3.1.1}$$

A proof of (3.1.1) is omitted; the easiest one involves comparing the coefficients of both sides, see Exercise 2.13.

For convenience, we take $Z(S_0) = 1$ and we use $Z(S_\infty)$ instead of $\sum_{n=0}^{\infty} Z(S_n)$. Then the left side of (3.1.1) may be denoted by $Z(S_\infty, f(x))$.

Now let

$$T(x) = \sum_{p=1}^{\infty} T_p x^p \tag{3.1.2}$$

be the generating function for rooted trees. Thus T_p is the number of rooted trees of order p. The rooted trees of order 4 or fewer are shown in Figure 3.1.1. Hence the first four terms of $T(x)$ are given by

$$T(x) = x + x^2 + 2x^3 + 4x^4 + \cdots. \tag{3.1.3}$$

The following result of Pólya [P8] can be used to calculate the coefficients of $T(x)$.

Theorem   The counting series $T(x)$ for rooted trees satisfies

$$T(x) = x \exp \left\{ \sum_{k=1}^{\infty} T(x^k)/k \right\}. \tag{3.1.4}$$

Proof   We shall first find the generating function which enumerates rooted trees in which the root has degree n. We observe that each of the latter trees corresponds in a natural way to a "combination with repetition" of n rooted trees. This correspondence is indicated for $n = 4$ in the next figure. More

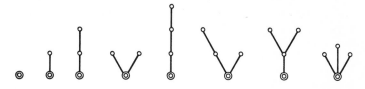

Figure 3.1.1

The smallest rooted trees.

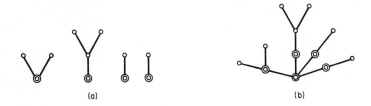

Figure 3.1.2

Four rooted trees and the corresponding tree whose root has degree 4.

specifically, given a collection of n rooted trees, a new rooted tree is formed by adding one new point and making it adjacent to each of the roots of the n given rooted trees. Clearly all trees whose roots have degree n can be formed in this manner. To find out how many there are, we consider the power group E^{S_n} with object set Y^X where E is the identity group, $X = \{1, \ldots, n\}$, and Y is the set of all rooted trees. Then each function in Y^X corresponds to an ordered n-tuple of rooted trees. We define the *weight* of each rooted tree in Y to be the number of points in the tree. Then $T(x)$ enumerates the elements of Y by weight and is called the "figure counting series" for Y. Thus the weight of each function in Y^X, as defined by (2.4.4), is the total number of points in the n rooted trees of the n-tuple to which the function corresponds.

Since S_n consists of all permutations of X, the orbits of the power group E^{S_n} correspond precisely to rooted trees whose root has degree n. Note that the weight of each orbit, which is the weight of any function in it, is just one less than the total number of points in the rooted tree to which the orbit corresponds. Therefore on applying PET with $A = S_n$ and $T(x)$ as the figure counting series, we have $Z(S_n, T(x))$ as the function counting series, and the coefficient of x^p in $Z(S_n, T(x))$ is the number of rooted trees of order $p + 1$ whose roots have degree n. Multiplication of $Z(S_n, T(x))$ by x corrects the weights so that the coefficient of x^p in $xZ(S_n, T(x))$ is the number of these trees with p points. Then on summing over all possible values of n, $T(x)$ itself is obtained:

$$T(x) = x \sum_{n=0}^{\infty} Z(S_n, T(x)). \qquad (3.1.5)$$

The proof is completed by applying the identity (3.1.1) for sums of cycle indexes to the right side of (3.1.5). **//**

It follows from this theorem that $T(x)$ is uniquely determined by the functional equation (3.1.4) because formula (1.2.8) provides a means of determining the coefficients of $T(x)$ inductively. To see this, let

$$\sum_{m=1}^{\infty} a_m x^m = \sum_{k=1}^{\infty} T(x^k)/k. \tag{3.1.6}$$

Therefore

$$a_m = m^{-1} \sum_{d|m} d T_d \tag{3.1.7}$$

and from formula (1.2.8) it follows as in Otter [O4] that

$$T_{p+1} = p^{-1} \sum_{k=1}^{p} k a_k T_{p-k+1}. \tag{3.1.8}$$

When (3.1.7) and (3.1.8) are combined, T_{p+1} is expressed in terms of $T_1, \ldots, T_p$:

$$T_{p+1} = p^{-1} \sum_{k=1}^{p} \left(\sum_{d|k} d T_d \right) T_{p-k+1}. \tag{3.1.9}$$

Earlier Cayley [C2] had found the following formula for $T(x)$ which is easily derived from (3.1.4) and vice versa:

$$T(x) = x \prod_{p=1}^{\infty} (1 - x^p)^{-T_p}. \tag{3.1.10}$$

To derive (3.1.4) from (3.1.10), it is sufficient to show that

$$\log(T(x)/x) = \sum_{k=1}^{\infty} T(x^k)/k. \tag{3.1.11}$$

This is accomplished in three steps by taking logarithms in (3.1.10), substituting the identity

$$\log(1 - x^p) = - \sum_{k=1}^{\infty} x^{pk}/k, \tag{3.1.12}$$

and interchanging the order of summation.

The number of rooted trees of order p has been determined for $p \le 26$ in Riordan [R15, p. 138] and for $p \le 39$ by A. J. Schwenk using (3.1.4). Here are the first few terms:

$$T(x) = x + x^2 + 2x^3 + 4x^4 + 9x^5 + 20x^6 + 48x^7$$
$$+ 115x^8 + 286x^9 + 719x^{10} + \cdots. \tag{3.1.13}$$

The coefficients of $T(x)$ for $p \le 26$ are found in Appendix I.

3.2 UNROOTED TREES

Now that all 1–1 functions on n objects can be counted as in Section 2.6, we need only the special case $n = 2$ to count unrooted trees. Let

$$t(x) = \sum_{p=1}^{\infty} t_p x^p \tag{3.2.1}$$

be the generating function for trees, so that t_p is the number of trees of order p. The neatest possible formula that expresses the series $t(x)$ for trees in terms of the series $T(x)$ for unrooted trees is provided by the main theorem of this section in formula (3.2.4). In addition to the counting theorem for 1–1 functions, the proof depends on a corollary of the dissimilarity characteristic theorem introduced next.

For any graph G, let p^* be the *number of dissimilar points*, i.e. the number of orbits of points determined by $\Gamma(G)$. The group of G also determines similarity classes of blocks. Let p_i^* be the number of dissimilar points in the ith class of blocks among the b^* dissimilar blocks. Then p^* and p_i^* are related by formula (3.2.2) of the next theorem. Otter [O4] first found this result for trees, but like so many theorems its proof is easier for the more general case [HN2]. Therefore we consider arbitrary blocks instead of just lines as in a tree.

Theorem (Dissimilarity characteristic theorem for graphs) For any graph G,

$$p^* - 1 = \sum_{i=1}^{b^*} (p_i^* - 1). \tag{3.2.2}$$

To illustrate the theorem, we consider the graph of order 18 in Figure 3.2.1. The points of this graph have been labeled so that similar points have the same labels. Thus $p^* = 4$. There are three classes of blocks, $b^* = 3$;

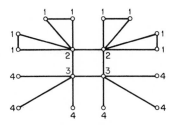

Figure 3.2.1

A graph with three dissimilar blocks.

the first consists of the bridges, the second is the 4-cycle, and the third contains the triangles. There are two dissimilar points in each class so that $p_1^* = p_2^* = p_3^* = 2$. Thus $3 = p^* - 1$ and $p_1^* - 1 + p_2^* - 1 + p_3^* - 1 = 3$ also.

Proof   The proof of the theorem is made by induction on number of classes of blocks. Let G be any graph. If there is just one class of blocks, $b^* = 1$, then $p^* = p_1^*$ and (3.2.2) obviously holds. Otherwise, consider any block of G that has exactly one cutpoint and suppose this block belongs to class number one. Delete from G the points of the members of this class except for the cutpoints. The graph G' so obtained has $b^* - 1$ classes of blocks and $p^* - (p_1^* - 1)$ classes of points. On applying the induction hypothesis to G', we have (3.2.2) for G.

To apply this theorem to trees, let p^* and q^* be the number of dissimilar points and lines, respectively, of any tree T. A line of T is called a *symmetry line* if its endpoints are similar, and s is the number of symmetry lines. The tree in Figure 3.2.2a has no symmetry lines, while that in (b) has one. The tree in (a) has $p^* = 6$ and each of its five classes of blocks has two dissimilar points. Hence $5 = p^* - 1 = \sum_{i=1}^{5} (2 - 1) = 5$. The tree in (b) has $p^* = 2$, and one class of blocks has two dissimilar points while the other corresponding to the symmetry line has only one. Hence $1 = p^* - 1 = (2 - 1) + (1 - 1) = 1$. The corollary as found by Otter then takes the following form.

Corollary   (Dissimilarity characteristic theorem for trees)   The number s of symmetry lines of any tree is 0 or 1 and

$$p^* - (q^* - s) = 1. \qquad (3.2.3)$$

Proof   We first observe that $s = 1$ if and only if the tree is bicentered and the two central points are similar; and otherwise $s = 0$ (see [H1, p. 35]). Note that q^* is the number of dissimilar blocks, so $b^* = q^*$. Furthermore $p_i^* = 2$ for each class of blocks (lines) other than symmetry lines, for which

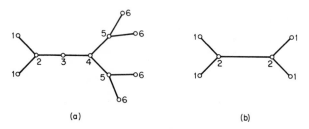

(a)   (b)

Figure 3.2.2

Trees with and without symmetry lines.

$p_i^* = 1$ by definition. Therefore the right side of (3.2.2) is $q^* - s$ and the proof is completed.   //

Now we are in a position to complete our derivation of Otter's elegant formula for $t(x)$.

Theorem The counting series $t(x)$ for trees is expressed in terms of the series $T(x)$ for rooted trees by

$$t(x) = T(x) - \tfrac{1}{2}(T^2(x) - T(x^2)). \tag{3.2.4}$$

Proof The first step of the proof is to sum (3.2.3) over all trees with exactly p points. The result is

$$\sum 1 = \sum p^* - \sum(q^* - s), \tag{3.2.5}$$

but $\sum 1 = t_p$ and $\sum p^*$ is T_p. Furthermore, $\sum(q^* - s)$ is the number L_p of trees with p points rooted at a line which is not a symmetry line. Therefore $t_p = T_p - L_p$; so if $L(x)$ is the counting series for trees rooted at a non-symmetry line, then

$$t(x) = T(x) - L(x). \tag{3.2.6}$$

At this point we can apply formula (2.6.1) of our theorem on 1–1 functions to express $L(x)$ in terms of $T(x)$. Note than any two *different* rooted trees determine a tree rooted at a nonsymmetry line and this correspondence illustrated in Figure 3.2.3 can be specified by joining the two roots by a distinguished line. Thus the trees counted by $L(x)$ can be interpreted as 2-subsets of the graphs enumerated by $T(x)$. Therefore we can apply (2.6.1) with $n = 2$ to obtain

$$L(x) = Z(A_2 - S_2, T(x)). \tag{3.2.7}$$

Now $Z(A_2 - S_2) = \tfrac{1}{2}(s_1^2 - s_2)$, and on setting $s_1 = T(x)$ and $s_2 = T(x^2)$, formulas (3.2.7) and (3.2.6) yield the Otter formula for $t(x)$.   //

Just as Cayley anticipated Pólya on $T(x)$ so he did Otter on $t(x)$. In fact it is a routine exercise to verify (3.2.4) from Cayley's original formula or from

Figure 3.2.3

Two rooted trees and the corresponding line-rooted tree.

Pólya's formula for t_p. Both Cayley and Pólya found involved expressions for t_p as the sum of t'_p and t''_p, the numbers of trees with 1 and 2 central points respectively. The verification is accomplished by forming $\sum t_p x^p$ and using the proverb "Every time you see a double summation sign interchange them."

The values of t_p for $p \leq 26$ have been computed using (3.2.4) by Riordan [R15, p. 138]. The first few terms of $t(x)$ are given here and the rest of these are in Appendix I:

$$t(x) = x + x^2 + x^3 + 2x^4 + 3x^5 + 6x^6 + 11x^7$$

$$+ 23x^8 + 47x^9 + 106x^{10} + \cdots. \qquad (3.2.8)$$

A *forest* is a graph whose components are trees. Now that trees have been counted, the enumeration of forests follows easily. Let the counting polynomial for forests with p points be

$$f_p(x) = \sum_{q=0}^{p-1} f_{p,q} x^q, \qquad (3.2.9)$$

where $f_{p,q}$ is the number of forests with p points and q lines. Then the generating function for forests is

$$f(x, y) = \sum_{p=1}^{\infty} y^p f_p(x). \qquad (3.2.10)$$

To derive formulas for $f_p(x)$ and $f(x, y)$, use is made of the counting series for trees. The formula in [H4] for $f(x, y)$ is obtained by the appropriate application of PET and the generalization of formula (3.1.1) for any function $g(x, y)$ of two variables. Thus the number of forests is expressed in terms of the number of trees by

$$1 + f(x, y) = \exp \sum_{n=1}^{\infty} \sum_{k=1}^{\infty} (t_k/n)(x^{k-1}y^k)^n. \qquad (3.2.11)$$

Using logarithms it is easily seen that this can also be expressed as

$$1 + f(x, y) = \prod_{k=1}^{\infty} (1 - x^{k-1}y^k)^{-t_k}, \qquad (3.2.12)$$

which resembles the form of Cayley's solution [C2] for the number of rooted trees. Now we give a more explicit formula [HP11] for $f_p(x)$ expressed in terms of the numbers t_k.

Theorem The counting polynomial for forests with p points is

$$f_p(x) = \sum_{(j)} \prod_{k=1}^{p} \binom{t_k + j_k - 1}{j_k} x^{(k-1)j_k}, \qquad (3.2.13)$$

and the sum is over all partitions (j) of p.

Proof Using the familiar identity for combinations with repetition (see [R15, p. 7]). We find that the number of forests consisting of exactly j_k trees, each of which has exactly k points, is the binomial coefficient

$$\binom{t_k + j_k - 1}{j_k}$$

Since each of these trees has $k - 1$ lines we have for each $q = 0$ to $p - 1$

$$f_{p,q} = \sum_{(j)} \prod_{k=1}^{p} \binom{t_k + j_k - 1}{j_k}. \qquad (3.2.14)$$

where the sum is over those partitions $(j) = (j_1, j_2, \ldots, j_p)$ of p such that

$$q = \sum_{k=1}^{p} (k - 1)j_k. \qquad (3.2.15)$$

The formula (3.2.13) for $f_p(x)$ may now be obtained by summing over *all* partitions of p. $\parallel$

For example, using the series $t(x)$ for trees and (3.2.13) with $p = 6$, one easily finds

$$f_6(x) = 1 + x + 2x^2 + 4x^3 + 6x^4 + 6x^5. \qquad (3.2.16)$$

On multiplying equation (3.2.13) by y^p and summing over all positive integers p, one can obtain (3.2.11) or (3.2.12) by straightforward manipulation.

3.3 TREES WITH SPECIFIED PROPERTIES

Many classes of trees can be enumerated by following the procedure in the previous sections of this chapter. Usually the generating function for the rooted variety is determined first by using PET. Then a "dissimilarity characteristic theorem" provides the means for expressing the series for unrooted trees in terms of that for rooted trees. In this section we shall consider several problems which can be treated successfully in this manner.

The first problem is to enumerate oriented trees. An *oriented tree* is a tree in which each line is assigned a unique direction. Let $r(x)$ and $R(x)$ be the counting series for oriented trees and rooted oriented trees respectively. All eight oriented trees of order 4 are shown in Figure 3.3.1, verifying that the coefficient of x^4 in $r(x)$ is 8.

The following result of [HP14] serves to determine the number of oriented trees.

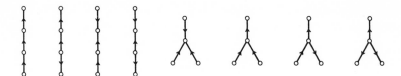

Figure 3.3.1

The oriented trees of order 4.

Theorem The counting series $r(x)$ and $R(x)$ for oriented trees and rooted oriented trees satisfy:

$$R(x) = x\left(\exp\left\{\sum_{k=1}^{\infty} R(x^k)/k\right\}\right)^2 \qquad (3.3.1)$$

and

$$r(x) = R(x) - R^2(x). \qquad (3.3.2)$$

Specifically, we find

$$r(x) = x + x^2 + 3x^3 + 8x^4 + 27x^5 + 91x^6 + \cdots. \qquad (3.3.3)$$

The first 21 coefficients of $r(x)$ and $R(x)$ are found in [R15, p. 138].

Proof We define a *planted tree* to be a rooted tree in which the root has degree 1 and we let $\bar{R}(x)$ be the counting series for planted, oriented trees. The r-subsets of these planted trees correspond to rooted oriented trees in which there are n arcs incident with the root (see Figure 3.3.2). On applying PET to the symmetric group S_n with $\bar{R}(x)/x$ as the figure counting series, we obtain $Z(S_n, \bar{R}(x)/x)$ as the function counting series in which the coefficient of x^{p-1} is the number of rooted, oriented trees of order p where the root is incident with n arcs. The use of $\bar{R}(x)/x$ as the figure counting series here in effect assigns a weight of zero to the roots of the planted trees, and hence to the roots of the rooted trees. On multiplying $Z(S_n, \bar{R}(x)/x)$ by x,

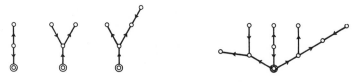

Figure 3.3.2

Three planted, oriented trees and their corresponding tree.

therefore, the proper adjustment is made. Then on summing over all non-negative integers n, $R(x)$ is expressed in terms of $\bar{R}(x)$ by

$$R(x) = xZ(S_\infty, \bar{R}(x)/x). \tag{3.3.4}$$

But for every rooted tree of order p we can construct two different planted trees of order $p + 1$ by adding a new arc directed either to or away from the root. Since all planted trees can be obtained uniquely in this way,

$$\bar{R}(x) = 2xR(x). \tag{3.3.5}$$

On substituting this series for $\bar{R}(x)$ in (3.3.4) and applying the identity (3.1.1), formula (3.3.1) of the theorem is verified.

To express $r(x)$ in terms of $R(x)$, we observe that oriented trees have no symmetry lines, $s = 0$. Hence the dissimilarity characteristic equation for these trees is simply

$$1 = p^* - q^*. \tag{3.3.6}$$

As in the case for ordinary trees, it follows from (3.3.6) that

$$r(x) = R(x) - L(x), \tag{3.3.7}$$

where $L(x)$ is the enumerator† of oriented trees rooted at an arc. These trees rooted at an arc correspond precisely to functions from the set $\{1, 2\}$ into the set of rooted, oriented trees. Given such a function f, the corresponding tree rooted at an arc is obtained by adding the root arc directed from the root of $f(1)$ to the root of $f(2)$. The enumerator of such functions for which the figure counting series is $R(x)$ is simply $R^2(x)$. Note that one obtains the same result by applying PET, namely $Z(E_2, R(x))$. Thus $L(x) = R^2(x)$ and the proof is completed on making this substitution in (3.3.7).   //

Next we shall consider *homeomorphically irreducible* trees, which have no points of degree 2. Those through order 8 are shown in the next figure. Let $h(x)$, $H(x)$, and $\bar{H}(x)$ be the counting series for homeomorphically irreducible trees, rooted trees, and planted trees respectively. The coefficients of these series can be calculated using the relations in the next theorem of [HP14].

---

† We already used $L(x)$ in the preceding section for trees rooted at a line not a symmetry line, and in Section 3.5, the same notation $L(x)$ will be used in a similar way for 2-trees. We hope the meaning will always be clear by context.

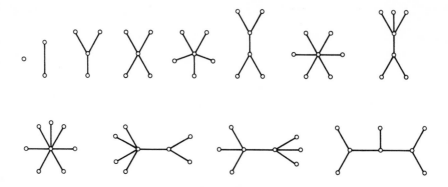

Figure 3.3.3

The smallest homeomorphically irreducible trees.

Theorem   The counting series $\bar{H}(x)$, $H(x)$, and $h(x)$ for homeomorphically irreducible trees satisfy:

$$\bar{H}(x) = \frac{x^2}{1 + x} \exp\left\{ \sum_{k=1}^{\infty} \frac{\bar{H}(x^k)}{kx^k} \right\} \tag{3.3.8}$$

$$H(x) = \frac{1 + x}{x}\bar{H}(x) - \frac{1}{2x}(\bar{H}^2(x) + \bar{H}(x^2)) \tag{3.3.9}$$

$$h(x) = \left(\frac{x - 1}{x}\right) H(x) + \left(\frac{1 + x}{x^2}\right)\bar{H}(x). \tag{3.3.10}$$

Explicitly,

$$h(x) = x + x^2 + x^4 + x^5 + 2x^6 + 2x^7 + 4x^8 + 5x^9$$
$$+ 10x^{10} + 14x^{11} + 26x^{12} + \cdots. \tag{3.3.11}$$

Proof   We begin by observing that n-subsets of planted trees correspond to planted trees in which the point adjacent to the root has degree n. This correspondence is indicated in Figure 3.3.4 where the roots of three planted trees are identified and a new point is introduced as the root of the new planted tree. With $\bar{H}(x)/x$ as the figure counting series, $Z(S_n, \bar{H}(x)/x)$ enumerates these planted trees, but the new root and the identified points have not yet been taken into account. The proper adjustment is made by multiplying by x^2. Then on summing over all $n \geq 2$ we again obtain $\bar{H}(x)$, but must add x^2 to allow for the planted tree of order 2:

$$\bar{H}(x) = x^2 + x^2 \sum_{n=2}^{\infty} Z(S_n, \bar{H}(x)/x). \tag{3.3.12}$$

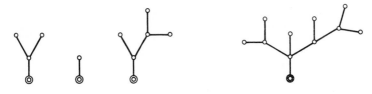

Figure 3.3.4

Three planted trees and the corresponding planted tree of order 11.

The identity (3.1.1) can now be applied to (3.3.12) to yield the first equation (3.3.8) of the theorem, from which the coefficients of $\bar{H}(x)$ can be calculated.

Next we verify (3.3.9) which expresses rooted trees in terms of planted trees. Now $Z(S_2, H(x))$ counts planted trees in which the point adjacent to the root has degree 3, and $x(H(x) - \bar{H}(x))$ counts planted trees in which the point adjacent to the root has degree 1 or greater than 3. Hence $\bar{H}(x)$ also satisfies

$$\bar{H}(x) = x(H(x) - \bar{H}(x)) + Z(S_2, \bar{H}(x)). \qquad (3.3.13)$$

On carrying out the substitution of $\bar{H}(x)$ in $Z(S_2)$ and solving for $H(x)$, the formula (3.3.9) is obtained.

Finally, we require the counting series $L(x)$ for homeomorphically irreducible trees rooted at an unsymmetric line, so that we can apply the dissimilarity characteristic theorem and express $h(x)$ in terms of $H(x)$ and $\bar{H}(x)$. To do this we observe that every pair of different planted trees with a total of k points corresponds to a tree of order $k + 2$ rooted at an unsymmetric line. This 1–1 correspondence is specified by joining the points adjacent to the roots of the two planted trees and deleting the roots as in Figure 3.3.5. Formula (2.6.1) of the theorem on 1–1 functions is again used to obtain $Z(A_2 - S_2, \bar{H}(x))$ as the enumerator of pairs of different planted trees. On division by x^2 the weights are properly adjusted, and then it follows from the dissimilarity characteristic equation (3.2.3) that

$$h(x) = H(x) - x^{-2}Z(A_2 - S_2, \bar{H}(x)). \qquad (3.3.14)$$

Figure 3.3.5

Two planted trees and the corresponding line-rooted tree.

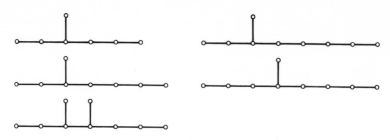

Figure 3.3.6

Small identity trees.

On substituting $\overline{H}(x)$ in $Z(A_2 - S_2)$, equation (3.3.10) is obtained and the proof is completed. $//$

Next we consider *identity trees*, whose automorphism group is the identity group. The identity trees of orders 7 through 9 are displayed in Figure 3.3.6. The only one of order less than 7 is the trivial tree.

The *absolute* $|T|$ of a rooted or line rooted tree T is the underlying unrooted tree with the same points and lines as T. It is clear that if $\Gamma(|T|)$ is the identity group, then so is $\Gamma(T)$, but not conversely. The small rooted identity trees are shown in Figure 3.3.7. Note that the groups of their absolutes are not necessarily the identity.

The following theorem of [HP14] relates the series for rooted and unrooted identity trees.

Theorem   Let $u(x)$ and $U(x)$ be the counting series for trees and rooted trees whose group is the identity. Then

$$U(x) = x \exp \sum_{k=1}^{\infty} (-1)^{k+1} U(x^k)/k \qquad (3.3.15)$$

$$u(x) = U(x) - \tfrac{1}{2}(U^2(x) + U(x^2)). \qquad (3.3.16)$$

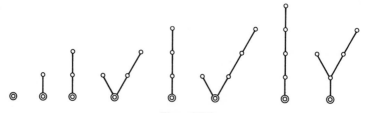

Figure 3.3.7

The small rooted identity trees.

Proof The verification of (3.3.15) is straightforward. From formula (2.6.1) in Pólya's theorem for 1–1 functions, we see that $xZ(A_n - S_n, U(x))$ is the counting series for rooted trees with trivial group and root degree $n \geq 1$. Summing over all n, (2.6.5) yields the formula for $U(x)$.

Since there are no symmetry lines in these trees, the dissimilarity characteristic equation is simply $1 = p^* - q^* = p - q$. Therefore to express $u(x)$ in terms of $U(x)$ we seek generating functions which enumerate the number of ways that identity trees *can be* rooted at points as well as lines. Specifically, let $U_1(x)$ and $U_2(x)$ be the respective counting series for rooted and line-rooted trees the group of whose absolute is the identity. It then follows that

$$u(x) = U_1(x) - U_2(x). \qquad (3.3.17)$$

In attempting to express $U_1(x)$ and $U_2(x)$ in terms of $u(x)$, the first step is to let $V_1(x)$ and $V_2(x)$ be the respective generating functions for rooted trees and line-rooted trees T which have the property that $\Gamma(T)$ is the identity group but $\Gamma(|T|)$ is not. Then we can write

$$U_1(x) = U(x) - V_1(x) \qquad (3.3.18)$$

and

$$U_2(x) = Z(A_2 - S_2, U(x)) - V_2(x). \qquad (3.3.19)$$

At this point, however, we observe that since

$$u(x) = U(x) - Z(A_2 - S_2, u(x)) + V_2(x) - V_1(x), \qquad (3.3.20)$$

we need only determine the difference between $V_1(x)$ and $V_2(x)$ in terms of $U(x)$. In particular, we now show that

$$V_1(x) - V_2(x) = U(x^2). \qquad (3.3.21)$$

We consider all trees T whose group is not the identity and then have two cases for the contributions of these trees to $V_1(x) - V_2(x)$.

Case 1   T has no symmetry line. We investigate how many rooted trees T' and line-rooted trees T'' with identity group have T as absolute. If there exist any such trees T' or T'', then $\Gamma(T)$ has exactly one element besides the identity, and this element must permute two branches at some point v_1 of T. Each of the two similar branches at v_1, considered as rooted trees, has the identity group. If each of these branches has $n + 1$ points, then there are exactly n rooted trees T' such that $|T'| = T$. Moreover, the line-rooted trees T'' obtained by rooting the n lines of one of these two branches also have $|T''| = T$. We conclude that for all these trees the number of rooted trees with identity group and absolute T equals the number of line-rooted trees with identity group and absolute T. Hence their contribution to $V_1(x) - V_2(x)$ is 0.

Case 2   T has a symmetry line. Here the order of $\Gamma(T)$ is at least 2. If there exist rooted trees T' or line-rooted trees T'' with identity group and absolute T, then $\Gamma(T)$ has order 2 and the nonidentity element permutes the central points of T. Therefore if T has $2n$ points, there are n rooted trees T' such that $|T'| = T$, and $n - 1$ line-rooted trees T'' such that $|T''| = T$. Thus for each tree T with a symmetry line and a group of order 2, the number of rooted trees with absolute T is one greater than the number of line-rooted trees with absolute T. Thus each such tree of order p contributes 1 to the coefficient of x^p in $V_1(x) - V_2(x)$.

Therefore we can conclude that $V_1(x) - V_2(x)$ is the generating function for trees with a symmetry line whose groups have order 2. But $U(x^2)$ enumerates these trees and hence (3.3.21) is verified. Now the entire proof is completed by substituting $-U(x^2)$ for $V_2(x) - V_1(x)$ in (3.3.20).   //

Formulas (3.3.15) and (3.3.16) of the theorem have been used to determine the following coefficients:

$$u(x) = x + x^7 + x^8 + 3x^9 + 6x^{10} + 15x^{11} + 29x^{12} + \cdots. \tag{3.3.22}$$

In addition to oriented, homeomorphically irreducible, and identity trees, formulas may be found in [HP14] for numerous other species including:

1.   trees with a given partition (or degree specification);
2.   trees with a given diameter;
3.   *directed trees*, in which each line is assigned one direction or both directions;
4.   *signed trees*, in which each line is assigned a plus or minus sign;
5.   trees of *strength s*, in which there are at most s lines between any pair of points;
6.   trees of given *weight*, in which integral weights are assigned to the points and the weight of a tree is the sum of the weights of its points.

All of these tree-counting problems are solved in a manner analogous to that used for the three solutions discussed above. Similarly, one can enumerate trees whose points have degree 1 or n with $n \geq 3$. In fact Pólya [P7] solved the latter problem for the case $n = 4$, thus determining the generating function for the saturated hydrocarbons, C_nH_{2n+2}.

Since trees can be embedded in the plane, we can ask for the number of *plane trees* of order p (see Figure 3.3.8). When a rooted tree is embedded in the plane, a cyclic order is induced on the lines incident with the root. It is shown in [HPT1] that the generating function $P(x)$ for rooted plane trees can be expressed in terms of the cycle index sum of the cyclic groups and the series $\bar{P}(x)$ for planted, plane trees. Then it is shown that the series

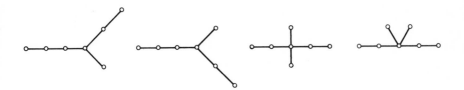

Figure 3.3.8

Four different plane trees of order 7.

$p(x)$ for plane trees is determined by $\bar{P}(x)$ and $P(x)$. Furthermore, the coefficients of $\bar{P}(x)$ can be determined in the explicit form of the first equation of the following theorem.

Theorem   If $\bar{P}(x)$, $P(x)$, and $p(x)$ are the counting series for planted, rooted, and ordinary plane trees, then

$$\bar{P}(x) = \sum_{n=1}^{\infty} \frac{1}{n}\binom{2n-2}{n-1} x^{n+1}, \qquad (3.3.23)$$

$$P(x) = x \sum_{n=0}^{\infty} Z(C_n, \bar{P}(x)/x), \qquad (3.3.24)$$

$$p(x) = P(x) - (1/2x^2)[\bar{P}^2(x) - \bar{P}(x^2)]. \qquad (3.3.25)$$

The first few terms of $p(x)$ are

$$p(x) = x + x^2 + x^3 + 2x^4 + 3x^5 + 6x^6 + 14x^7 + \cdots. \qquad (3.3.26)$$

Curiously, $\bar{P}(x)$ also counts the planted plane trees in which each point has degree 1 or 3. That is, the number of planted plane trees of order p is also the number of planted plane trees with $p - 2$ points of degree 3 and $p - 1$ points of degree 1. This fact is illustrated for $p = 5$ in the next two figures. The dual form of this observation asserts that $\bar{P}(x)$ counts the number

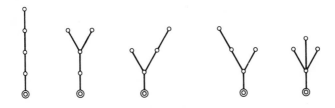

Figure 3.3.9

The five, planted, plane trees of order 5.

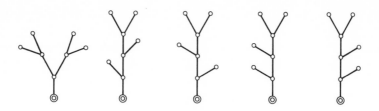

Figure 3.3.10

The five planted, plane trees whose points have degree 1 or 3.

of ways of subdividing a convex n-gon rooted at an oriented line into triangular faces by means of diagonals. The five pentagons corresponding to the trees of Figure 3.3.10 are displayed in Figure 3.3.11. Brown [B3] points out that the result has been discovered many times, and traces it back to Euler [E1].

The coefficients of $\bar{P}(x)$, usually called the Catalan numbers, also count the number of nonassociative products of n terms. The connection of trees with such parenthesizing schemes is beautifully developed by Comtet in [C5, p. 64].

The list of solved tree-counting problems is extensive and therefore many more of them are introduced in the exercises.

3.4 TREELIKE GRAPHS

The techniques of this chapter can be adapted to obtain generating functions for many classes of graphs which either resemble trees or contain trees as induced subgraphs. We shall consider in some detail the problems of counting unicyclic graphs, functional digraphs, block-cutpoint-trees, block graphs, and cacti.

We shall first count unicyclic graphs because the approach used to enumerate them can be specialized to count functional digraphs. A unicyclic graph is connected and has just one cycle. If G is unicyclic and its cycle has length n, then G may be regarded as having a rooted tree, possibly the

Figure 3.3.11

The five triangulated pentagons with an oriented boundary line.

trivial one, attached to each of the n points of its cycle. Therefore let Y be the set of rooted trees with counting series $T(x)$ found in equation (3.1.4). If the power group E^{D_n} has object set Y^X, then the orbits of functions in Y^X correspond precisely to unicyclic graphs. Hence PET gives the next result.

Theorem The counting series $U_n(x)$ for unicyclic graphs whose cycle has length n is given by

$$U_n(x) = Z(D_n, T(x)). \tag{3.4.1}$$

See [R15, p. 150] for the coefficients of x^k in $U_n(x)$ with n and $k \leq 10$.

A digraph is *functional* if every point has outdegree 1. The concept of a functional digraph arises in a psychological context in the study of the structure of a group of people in which each member extends exactly one invitation to another member. Our object now is to find a generating function whose coefficients give the number of isomorphically distinct functional digraphs with a given number of points [H10]. It will be seen that these digraphs correspond to functions which are fixed-point free. Davis [D1] has found an explicit formula for the number of types of functions on a finite set (see also [P2]). His methods may be readily used to solve this variation of the problem. However, in the process of deriving this generating function, we find certain structural properties of functional digraphs which are of independent interest. In particular, a functional digraph is constructible from directed cycles and rooted trees.

If Z is a directed cycle of a functional digraph D, then by $D - Z$ we mean the digraph obtained from D on removing all the lines of Z. Recall that a tree to the point u is obtained from a rooted tree with root u on orienting each of its lines so that it is directed toward u. Now we are ready to characterize functional digraphs, the proof can be found in [HNC1, p. 325].

Theorem A digraph D is functional if and only if each of its weak components consist of exactly one directed cycle Z and for each point u of Z, the weak component $R(u)$ of $D - Z$ which contains u is a tree to the point u.

It follows from PET and this characterization theorem that the counting series $v(x)$ for functional digraphs (with n_k cycles of length k) is given by

$$v(x) = \sum \prod_{k=2}^{\infty} Z(S_{n_k}, Z(C_k, T(x))), \tag{3.4.2}$$

where the sum is over each $n_k = 0$ to ∞. On interchanging the sum and product symbols, we obtain

$$v(x) + 1 = \prod_{k=2}^{\infty} Z(S_\infty, Z(C_k, T(x))). \tag{3.4.3}$$

Using formula (3.1.1) for summing cycle indexes of the symmetric groups, we obtain the next formula.

Theorem   The counting series $v(x)$ for functional digraphs is given by

$$v(x) + 1 = \exp \sum_{n=1}^{\infty} (1/n) \sum_{k=2}^{\infty} Z(C_k, T(x^n)). \qquad (3.4.4)$$

By clever algebraic manipulations, Read [R3] reduced this formula to

$$v(x) + 1 = \frac{x}{T(x)} \prod_{n=1}^{\infty} (1 - T(x^n))^{-1}. \qquad (3.4.5)$$

The only difference between a function and a functional digraph is that the latter has no loops, or fixed points. Thus $v(x)$ enumerates (the isomorphism types of) functions that are fixed-point free. Only a slight modification of (3.4.3) is required to enumerate the total number of functions:

$$\prod_{k=1}^{\infty} Z(S_\infty, Z(C_k, T(x))). \qquad (3.4.6)$$

In this formula the coefficient of x^n is the formula "fcn(n)" of Davis [D1]. Read [R3] calculated the numbers of functional digraphs and functions displayed in Table 3.4.1.

TABLE 3.4.1

n	1	2	3	4	5	6	7	8	9	10	11
Functional digraphs	0	1	2	6	13	40	100	291	797	2 273	6 389
Functions	1	3	7	19	47	130	343	951	2 615	7 318	20 491

It has often been observed that a connected graph with quite a few cutpoints bears a resemblance to a tree. We now make this notion explicit by associating with every connected graph G a tree $bc(G)$ which reflects this resemblance, [HP16]. The *block-cutpoint-tree* $bc(G)$ is the graph whose set of points is the union of the set of blocks and the set of cutpoints of G, with two points adjacent if one corresponds to a block of G, and the other to a cutpoint of G in that block. It is easy to show that if G is connected, then $bc(G)$ is, indeed, a tree.

We now define a *bc-tree* as a bicolored tree in which every endpoint has the same color, say blue while the other color is coral. Thus the distance between any two endpoints is even. It then follows, see [H1, p. 36] that every *bc*-tree is the block-cutpoint tree of a connected graph and conversely. Therefore, to enumerate block-cutpoint trees, we need only count *bc*-trees. Let

$$t(x, y) = \sum_{m=1, n=0}^{\infty} t_{m,n} x^m y^n, \qquad (3.4.7)$$

where $t_{m,n}$ is the number of *bc*-trees with m blue points and n coral points. Similarly let $T(x, y)$, $T_B(x, y)$, $T_C(x, y)$ be the generating series for rooted *bc*-trees, *bc*-trees rooted at a blue point, and *bc*-trees rooted at a coral point, respectively.

Theorem The counting series for *bc*-trees satisfy:

$$T_C(x, y) = y(Z(S_\infty, T_B(x, y)) - T_B(x, y) - 1), \qquad (3.4.8)$$

$$T_B(x, y) = xZ(S_\infty, T_C(x, y)) + yT_B(x, y), \qquad (3.4.9)$$

$$T(x, y) = T_B(x, y) + T_C(x, y), \qquad (3.4.10)$$

$$t(x, y) = T(x, y) - T_B(x, y)(T_C(x, y) + yT_B(x, y)). \qquad (3.4.11)$$

The proof may be found in [HP16]. Explicitly,

$$\begin{aligned} t(x, y) = x + x^2 y + x^3(y + y^2) + x^4(y + y^2 + 2y^3) \\ + x^5(y + 2y^2 + 3y^3 + 3y^4) + \cdots. \end{aligned} \qquad (3.4.12)$$

The *block graph*, denoted $B(G)$, of a given graph G has as its points the blocks of G and two points are adjacent if the corresponding blocks have a point in common. Norman [N1] obtained generating functions for connected graphs in which every block is complete. These are shown to be block graphs in [H1, p. 30]. In light of this correspondence the formulas of the previous theorem can be used, following [HP16] to count block graphs.

Corollary The series $\bar{B}(x)$ and $B(x)$ that enumerate connected rooted and unrooted block graphs satisfy

$$\bar{B}(x) = T_B(x, 1) = x + x^2 + 3x^3 + 8x^4 + 25x^5 + \cdots \qquad (3.4.13)$$

$$B(x) = t(x, 1) = x + x^2 + 2x^3 + 4x^4 + 9x^5 + \cdots. \qquad (3.4.14)$$

The first four coefficients of (3.4.14) are verified in Figure 3.4.1.

A *cactus* is a connected graph in which no line lies on more than one cycle. These graphs were formerly called "Husimi trees" and their definition[†]

[†] This term received much criticism because Husimi trees are not necessarily trees.

Figure 3.4.1

The smallest block graphs.

was given by Uhlenbeck [UF1] and Riddell [R4] following a paper by Husimi [H16] on the cluster integrals in the theory of condensation in statistical mechanics. To enumerate them, we require an appropriate dissimilarity characteristic theorem [HN1], but here we shall only illustrate its use by counting *triangular cacti* [HN1] in which every line is in a triangle.

For a given cactus H, we denote by p^*, q^*, and r^* the number of dissimilar points, lines, and cycles respectively, and by s the number of symmetry lines not in a cycle. Let C be any cycle of H which has $n \geq 3$ points. Then $A = \Gamma(H)|_C$ is the group of H restricted to C and hence is a subgroup of D_n. Now suppose n is even and $|A| = 2$. Then there are two possibilities for the nontrivial element α of A. Either α fixes two points of C or it fixes two lines of C. In the first case, C is called a *type-1 cycle* and in the second, a *type-2 cycle*. Let r_i^* be the number of similarity classes in H of type-i cycles for $i = 1$ and 2. Now the theorem can be stated as follows (note that for all H we have the special Euler–Poincaré formula $1 = p - q + r$).

Theorem (Dissimilarity characteristic for cacti) The classes of points, lines and cycles for any cactus satisfy

$$1 = p^* - (q^* - s) + (r^* - r_1^* + r_2^*). \tag{3.4.15}$$

The details of the proof may be found in [HN1].

For brevity, we call a triangular cactus a $\triangle$-*cactus*. By specializing (3.4.15) to $\triangle$-cacti, we find

$$1 = p^* - q^* + r^*. \tag{3.4.16}$$

Therefore to enumerate these, it is necessary to find the generating function for $\triangle$-cacti which are rooted at a point, at a line, and at a triangle. For convenience, we adopt the convention that a single point is a $\triangle$-cactus. Let $D(x)$ be the generating function for these graphs which are rooted at a point, i.e., the coefficient of x^n is the number of trees with n triangles. If the root point has degree 2, the series is just $xZ(S_2, D(x))$, hence

$$D(x) = Z(S_\infty, xZ(S_2, D(x))). \tag{3.4.17}$$

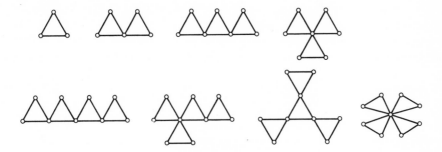

Figure 3.4.2

Small triangular cacti.

Then the series for these $\triangle$-cacti which are rooted at a line is $xD(x)Z(S_2, D(x))$ and for those rooted at a triangle $xZ(S_3, D(x))$. Now by summing formula (3.4.16) over all these we have an expression (3.4.19) for $d(x)$, the series that counts $\triangle$-cacti, in terms of the series $D(x)$ for rooted $\triangle$-cacti.

Theorem The series $D(x)$ and $d(x)$ for rooted and unrooted triangular cacti satisfy:

$$D(x) = \exp \sum_{k=1}^{\infty} \frac{x^k}{2k}(D^2(x^k) + D(x^{2k})), \tag{3.4.18}$$

$$d(x) = D(x) - \frac{x}{3}(D^3(x) - D(x^3)). \tag{3.4.19}$$

In particular, from (3.4.18) we find

$$D(x) = 1 + x + 2x^2 + 5x^3 + 13x^4 + 37x^5 + 111x^6$$
$$+ 345x^7 + 1105x^8 + \cdots. \tag{3.4.20}$$

From (3.4.19) it follows that

$$d(x) = 1 + x + x^2 + 2x^3 + 4x^4 + 8x^5 + 19x^6$$
$$+ 48x^7 + 126x^8 + \cdots, \tag{3.4.21}$$

and the first few coefficients are seen to agree with Figure 3.4.2. For a complete discussion of cacti consisting only of quadrilaterals, see [HU1].

3.5 TWO-TREES

In this section some higher dimensional concepts corresponding to trees are studied. In order to enumerate the two-dimensional structures so obtained,

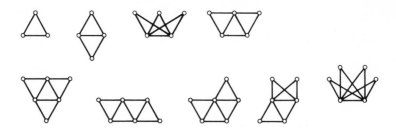

Figure 3.5.1

The graphs of the small 2-trees.

called 2-trees, a dissimilarity characteristic theory is investigated and Pólya's enumeration theorems are applied. Our methods can be specialized to count those 2-trees which are embeddable in the plane, thus providing a new approach to the old problem of determining the number of triangulations of a polygon.

In [HP12] we defined an *n-plex* as an *n*-dimensional simplicial complex in which every *k*-simplex with $k < n$ is contained in an *n*-simplex. We will only be concerned with 2-plexes, and for convenience 0-simplexes, 1-simplexes, and 2-simplexes are called *points*, *lines*, and *cells* respectively. The *two-dimensional trees*, also called *2-trees* can now be defined inductively. The 2-plex with three points is a 2-tree and a 2-tree with $p + 1$ points is obtained from a 2-tree with *p* points by adjoining a new point *w* adjacent to each of two adjacent points *u* and *v* together with the accompanying cell $\{u, v, w\}$. The definition of a *k*-tree for $k > 2$ is similar. For purposes of enumerating 2-trees, one need only consider their underlying graphs or 1-skeletons, which are shown in Figure 3.5.1 for $p \leq 6$.

By the *number of dissimilar points* p^* of a 2-tree we mean as usual the number of orbits of points; analogous definitions are made for the number q^* of dissimilar lines and r^* for cells.

Theorem (Dissimilarity characteristic for 2-trees) For any 2-tree with q^* dissimilar lines, q^* dissimilar cells, s_1 cells with two similar lines, s_2 cells with all three lines similar, and $s = s_1 + 2s_2$,

$$q^* + s - 2q^* = 1. \tag{3.5.1}$$

Now we proceed to develop the generating functions for 2-trees. Let t_n be the number† of 2-trees with *n* cells. The counting series for 2-trees is

---

† This is the same t_n notation as used earlier for trees and earlier yet for labeled trees, but we have run out of letters and hope that this will not cause too much confusion.

denoted by

$$t(x) = \sum_{n=1} t_n x^n. \tag{3.5.2}$$

In order to derive formulas for t_n, we will make use of the corresponding series for various kinds of rooted 2-trees. First let $M_1(x)$ and $N_1(x)$ be the series for 2-trees rooted at a symmetric and an unsymmetric end-line respectively. Further, let $M(x)$ and $N(x)$ be the series for 2-trees rooted at any symmetric and any unsymmetric line respectively. The following two equations express $M_1(x)$ and $N_1(x)$ in terms of $M(x)$ and $N(x)$:

$$M_1(x) = x(1 + M(x^2) + 2N(x^2)), \tag{3.5.3}$$

$$N_1(x) = xZ(A_2 - S_2, 1 + M(x) + 2N(x)). \tag{3.5.4}$$

Next we express $M(x)$ in terms of $M_1(x)$ and $N_1(x)$:

$$M(x) = \sum_{n=1}^{\infty} Z(S_n, M_1(x) + N_1(x^2)). \tag{3.5.5}$$

Using the identity (3.1.1), equation (3.5.5) may be written

$$1 + M(x) = \exp\left\{ \sum_{n=1}^{\infty} (1/n)[M_1(x^n) + N_1(x^{2n})] \right\}. \tag{3.5.6}$$

Now note that the counting series for 2-trees rooted at an oriented line is simply $M(x) + 2N(x)$. From this observation we have

$$M(x) + 2N(x) = \sum_{n=1}^{\infty} Z(S_n, M_1(x) + 2N_1(x)). \tag{3.5.7}$$

Again using the identity (3.1.1), we may write (3.5.7) as

$$1 + M(x) + 2N(x) = \exp\left\{ \sum_{n=1}^{\infty} (1/n)[M_1(x^n) + 2N_1(x^n)] \right\}. \tag{3.5.8}$$

Thus equations (3.5.6) and (3.5.8) may be used to solve for $N(x)$ in terms of $M_1(x)$ and $N_1(x)$. Now using all four formulas (3.5.3), (3.5.4), (3.5.6) and (3.5.8), the coefficients of $M(x)$ and $N(x)$ can be calculated. For the first few terms we have

$$M(x) = x + x^2 + 2x^3 + 3x^4 + 6x^5 + \cdots, \tag{3.5.9}$$

$$N(x) = x^2 + 4x^3 + 18x^4 + 77x^5 + \cdots. \tag{3.5.10}$$

The series for 2-trees rooted at a line is denoted $L(x)$ and since $L(x) = M(x) + N(x)$, we have immediately

$$L(x) = x + 2x^2 + 6x^3 + 21x^4 + 83x^5 + \cdots \tag{3.5.11}$$

We denote the series for 2-trees rooted at a cell (triangle) by $\triangle(x)$. It can be shown that

$$\triangle(x) = xZ(S_3, 1 + M(x) + 2N(x)) - xN(x)(1 + M(x^2) + 2N(x^2)). \quad (3.5.12)$$

Having expressed $\triangle(x)$ in terms of $M(x)$ and $N(x)$, we substitute (3.5.9) and (3.5.10) in equation (3.5.12) to obtain

$$\triangle(x) = x + x^2 + 3x^3 + 10x^4 + 39x^5 + \cdots. \quad (3.5.13)$$

Following the notation of (3.5.1), the Dissimilarity Characteristic Theorem for 2-trees, we denote by $s_1(x)$ the counting series for 2-trees rooted at a cell with two similar lines. Similarly, $s_2(x)$ is the series for 2-trees rooted at a cell with all three lines similar. These two series are readily expressed as functions of $M_1(x)$, $M(x)$, and $N(x)$:

$$s_1(x) = M_1(x)(1 + M(x) - x(1 + M(x^3))), \quad (3.5.14)$$

$$s_2(x) = x(1 + M(x^3) + N(x^3)). \quad (3.5.15)$$

Making the appropriate substitutions, we obtain

$$s_1(x) = x^2 + 2x^3 + 2x^4 + 7x^5 + \cdots, \quad (3.5.16)$$

$$s_2(x) = x + x^4 + 2x^7 + 6x^{10} + \cdots. \quad (3.5.17)$$

In order to express the formula for $t(x)$, the series for 2-trees, we use the Dissimilarity Characteristic Theorem (3.5.1) in the same manner as was done by Otter for the enumeration of trees.

Theorem   (Enumeration Theorem for 2-trees)   The counting series for 2-trees is

$$t(x) = L(x) + s_1(x) + 2s_2(x) - 2\triangle(x). \quad (3.5.18)$$

Substituting equations (3.5.11), (3.5.13), (3.5.16), and (3.5.17) into equation (3.5.18) gives

$$t(x) = x + x^2 + 2x^3 + 5x^4 + 12x^5 + \cdots. \quad (3.5.19)$$

Note that this theorem can be used to count 2-trees with specified properties provided that formulas for $L(x)$, $s_1(x)$, and $\triangle(x)$ are found for 2-trees with these properties.

By a *triangulation of a polygon* we mean a graph obtained from a regular n-gon by adding nonintersecting chords until every interior region is a triangle. Obviously $n - 3$ chords are required and $n - 2$ triangles are obtained. Generating functions for the number of different triangulations of the n-gon, i.e., those not isomorphic as graphs, have been found by

Brown [B2], but our purpose here is to present an entirely different approach toward finding such a generating function. We alter the formulation of the problem into a statement involving two-dimensional simplicial complexes by observing that triangulations of a polygon correspond precisely with planar 2-trees. We then proceed to enumerate the latter by the same methods used to count 2-trees.

To illustrate the configurations being counted, we show in Figure 3.5.2 the unique triangulations of a triangle, a quadrilateral, and a pentagon, and the three different triangulations of a hexagon. Note that these graphs are not taken as rooted or labeled in any way. Observe also the correspondence between these and the planar 2-trees with one, two, three, and four cells.

The enumeration of planar 2-trees can be accomplished by using almost all of the formulas that have already been developed for 2-trees. Therefore we alter the notation used for 2-trees only slightly by writing a bar to indicate the generating functions for planar 2-trees.

Thus let $\overline{M}_1(x)$ and $\overline{N}_1(x)$ be the series for planar 2-trees rooted at a symmetric and an unsymmetric end-line respectively. Then the following two formulas (compare (3.5.3) and (3.5.4)) specify the relationship between $\overline{M}_1(x)$ and $\overline{N}_1(x)$:

$$\overline{M}_1(x) = x(1 + \overline{M}_1(x^2) + 2\overline{N}_1(x^2)), \tag{3.5.20}$$

$$\overline{N}_1(x) = xZ(A_2 - S_2, 1 + \overline{M}_1(x) + 2\overline{N}_1(x)). \tag{3.5.21}$$

These two equations can be used to obtain the coefficients in the two series $\overline{M}_1(x)$ and $\overline{N}_1(x)$. However, as noted in the enumeration of plane trees, a formula due to Euler (see Figure 3.3.11 and equation (3.3.23)) shows that

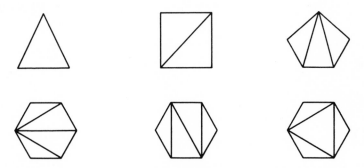

Figure 3.5.2

The triangulations of the n-gon, n = 3 to 6.

the number of triangulations of an $(n + 2)$-gon which is rooted by orienting one of its boundary edges is

$$f_n = \frac{2(2n - 1)!}{(n - 1)!(n + 1)!} = \frac{1}{n}\binom{2n}{n - 1}. \tag{3.5.22}$$

Hence it follows that

$$\overline{M}_1(x) + 2\overline{N}_1(x) = \sum_{n=1}^{\infty} \frac{2(2n - 1)!}{(n - 1)!(n + 1)!} x^n. \tag{3.5.23}$$

Now from Euler's formula (3.5.22) and equation (3.5.20) for $\overline{M}_1(x)$, we have

$$\overline{M}_1(x) = x + \sum_{n=1}^{\infty} \frac{2(2n - 1)!}{(n - 1)!(n - 1)!} x^{2n + 1}. \tag{3.5.24}$$

The first few terms of $\overline{M}_1(x)$ and $\overline{N}_1(x)$ are

$$\overline{M}_1(x) = x + x^3 + 2x^5 + 5x^7 + 14x^9 + \cdots. \tag{3.5.25}$$

$$\overline{N}_1(x) = x^2 + 2x^3 + 7x^4 + 20x^5 + 66x^6 + 212x^7 + 715x^8 + \cdots. \tag{3.5.26}$$

The series for planar 2-trees rooted at a line is denoted $\overline{L}(x)$ and can be expressed in terms of $\overline{M}_1(x)$ and $\overline{N}_1(x)$:

$$\overline{L}(x) = Z(S_2, 1 + \overline{M}_1(x) + \overline{N}_1(x)) + Z(S_2, \overline{N}_1(x)) - 1. \tag{3.5.27}$$

Substitution of (3.5.25) and (3.5.26) in equation (3.5.27) for $\overline{L}(x)$ yields

$$\overline{L}(x) = x + 2x^2 + 4x^3 + 12x^4 + 34x^5 + 111x^6 + 360x^7 \\ + 1226x^8 + \cdots. \tag{3.5.28}$$

From this point on, since the equations and procedures are virtually the same for planar 2-trees as for 2-trees, we will simply list the formulas for $\overline{\triangle}(x)$, $\bar{s}_1(x)$, $\bar{s}_2(x)$, and $\bar{t}(x)$:

$$\overline{\triangle}(x) = xZ(S_3, 1 + \overline{M}_1(x) + 2\overline{N}_1(x)) - x\overline{N}_1(x)(1 + \overline{M}_1(x^2) \\ + 2\overline{N}_1(x^2)), \tag{3.5.29}$$

$$\bar{s}_1(x) = \overline{M}_1(x)(1 + \overline{M}_1(x)) - x(1 + \overline{M}_1(x^3)), \tag{3.5.30}$$

$$\bar{s}_2(x) = x\overline{N}_1(x^3) + x(1 + \overline{M}_1(x^3)), \tag{3.5.31}$$

and as before

$$\bar{t}(x) = \overline{L}(x) + \bar{s}_1(x) + 2\bar{s}_2(x) - 2\overline{\triangle}(x), \tag{3.5.32}$$

which is obtained by barring equation (3.5.18), the Enumeration Theorem for 2-trees.

TABLE 3.5.1

PLANAR 2-TREES

n	t_n	n	t_n
1	1	13	24 834
2	1	14	83 898
3	1	15	285 357
4	3	16	1 046 609
5	4	17	3 412 420
6	12	18	11 944 614
7	27	19	42 080 170
8	82	20	149 197 152
9	228	21	532 883 768
10	733	22	1 905 930 975
11	2 282	23	6 861 221 666
12	7 528		

Substituting the calculations (3.5.25) and (3.5.26) for $\overline{M}_1(x)$ and $\overline{N}_1(x)$ in these formulas gives the series $\bar{t}(x)$ for planar 2-trees. In an unpublished work, R. K. Guy (to wile away his time in Singapore) used pencil, paper, and a desk calculator to obtain the first 23 coefficients of $\bar{t}(x)$ shown in Table 3.5.1.

EXERCISES

3.1 The *partition* of a tree is the sequence of nonnegative integers $(a_1, a_2, a_3, \ldots)$ where a_m is the number of points of degree m. Trees with a given partition. (Harary and Prins [HP14])

3.2 The *diameter* of a tree is the length of a longest path. Trees with a given diameter. (Harary and Prins [HP14])

3.3 A *weighted tree* is a tree to each of whose points is assigned a positive integer called its *weight*. The *weight* of a tree is the sum of the weights of its points. Trees with p points and weight m. (Harary and Prins [HP14])

3.4 A *signed tree* is one in which each line is assigned a plus or minus sign. Signed trees. (Harary and Prins [HP14])

3.5 A tree has *strength* s if at least one pair of points is jointed by s lines, but no two points are joined by more than s lines. Trees of strength $\leq s$. (Harary and Prins [HP14])

3.6 (a) Connected, (b) unilateral, (c) strong functional digraphs. (Harary [H10])

3.7 Cacti in which every block is a quadrilateral.

(Harary and Uhlenbeck [HU1])

3.8 Triangulations of a polygon with n interior points. (Brown [B2])

3.9 Trees with a given number of endpoints.

3.10 The generating function $U(x)$ for rooted identity trees satisfies

$$U(x) = x \prod_{n=1}^{\infty} (1 + x^n)^{U_n}.$$

3.11 Trees whose points are all of degree 1 or 4 (the saturated hydrocarbons).

(Pólya [P7])

3.12 A rooted tree has *height* k if k is the distance from the root to a farthest point. Rooted trees with given height. (Riordan [R16])

3.13 Rooted plane trees:

$$\frac{1}{2q} \sum_{d|q} \varphi\left(\frac{q}{d}\right)\binom{2d}{d}.$$

(Walkup [W1])

3.14 Forests with q lines and no isolated points.

Chapter 4 | GRAPHS

A solution to the fascinating problem of determining the number of graphs of order p was apparently first published in 1927. The author of the remarkable paper [R10] which contained this intriguing result was J. H. Redfield.† This pioneering paper went virtually unnoticed for about thirty years, but in the meantime the problem was tackled successfully and independently by several mathematicians including R. L. Davis [D1], A. M. Gleason, S. Golomb, D. Slepian, and of course, G. Pólya. As early as

---

† A letter dated 19 Dec. 1963 from C. Oakley to F. Harary reads in full:

Howard Redfield was a graduate of Haverford College in the Class of 1899. He was a man of very broad interests and we do not have a continuous record of his doings. Directly after leaving college, he worked as a civil engineer. In college he took a lot of languages and mathematics. (There was no major department in those days.) After graduating from Haverford with a B.S. degree, he took a S.B. degree in M.I.T. and a M.A. and Ph.D. (mathematics) at Harvard. During the year 1907–1908, he studied romance philology at the University of Paris. In 1908–1909, he was an instructor in mathematics at Worcester Polytechnic Institute, Worcester, Massachusetts. In 1910–1911 he taught French at Swarthmore College

81

1937 when his enumeration theorem appeared, Pólya was able to apply it to the problem of determining the generating functions that counted graphs according to the number of points and lines present. A letter from G. Pólya to F. Harary in 1951 contained formula (4.1.9) below with $1 + x$ substituted in it and illustrated it for $p = 4$. But it was not until the appearance of the paper [H4] in 1955 that the details were published.

We have already seen in Chapter 3 how Pólya's theorem can be applied to counting problems involving trees and treelike structures. This chapter presents further evidence of the wide applicability of his method. We shall enumerate graphs, rooted graphs, connected graphs, bicolored graphs, locally restricted graphs, symmetric graphs, boolean functions, and eulerian graphs.

4.1 GRAPHS

Pólya's efficient method for counting graphs requires the construction of a permutation group whose orbits correspond precisely to isomorphism classes of labeled graphs with p points and q lines. On deriving an explicit formula for the appropriate cycle index, an application of PET with figure counting series $1 + x$ determines the counting polynomial which has as the coefficient of x^q, the number of (p, q) graphs.

More specifically, let

$$g_p(x) = \sum_{q=0}^{m} g_{p,q} x^q \tag{4.1.1}$$

---

and from 1912–1914 he was an assistant professor of romance languages at Princeton University. From 1916 onward until his death in 1944, he was a practicing civil engineer in Wayne, Pennsylvania.

I knew him from about 1938–1944. Indeed in 1940 he came to Haverford College and gave us some lectures on "Electronic Digital Computers" (this was slightly before Eckert–Mauchly). Knowing him as I did in those later years, I could well understand how he would not make a great teacher. He was completely off in the clouds at all times. He never looked at you, he spoke softly with his eyes on the floor, he worked with his back to you and wrote on the board. His board work, however, was impeccable. It could have been photographed and printed by photo offset it was so perfect.

He came to Haverford to talk to our math club many times and always had something new to say. He was so modest that you never knew whether what he was doing was his own or somebody else's.

This is about all I know of him except for the fact that he has a very distinguished brother (Alfred Redfield) at Woods Hole, Massachusetts.

where $m = \binom{p}{2}$ and $g_{p,q}$ is the number of (p, q) graphs. For example, a glance at Figure 1.1.2 confirms that

$$g_4(x) = 1 + x + 2x^2 + 3x^3 + 2x^4 + x^5 + x^6. \qquad (4.1.2)$$

Note that the coefficients of these polynomials are always "end-symmetric" since the number of (p, q) graphs equals the number of $(p, m - q)$ graphs by complementation.

Since PET enumerates orbits of functions, we shall first provide a natural correspondence between graphs and functions. Let $X = \{1, \ldots, p\}$, while $X^{(2)}$ denotes all 2-subsets of X. Then with $Y = \{0, 1\}$, the functions from $X^{(2)}$ into Y represent labeled graphs of order p. Each function f corresponds to that graph $G(f)$ with point set X in which i and j are adjacent if and only if $f\{i, j\} = 1$. Thus two functions f and h represent the same graph if there is a permutation α of X such that whenever i and j are adjacent in $G(f)$, then αi and αj are adjacent in $G(h)$. Therefore $G(f)$ and $G(h)$ are isomorphic if and only if for some permutation α acting on X,

$$f\{i, j\} = h\{\alpha i, \alpha j\} \qquad (4.1.3)$$

for all $\{i, j\}$ in $X^{(2)}$. This equation suggests the following unary operation on permutation groups which leads to the group required for the enumeration of graphs.

Let A be a permutation group with object set $X = \{1, 2, \ldots, p\}$. The *pair group of* A, denoted $A^{(2)}$, is the permutation group induced by A which acts on $X^{(2)}$. Specifically, each permutation α in A induces a permutation α' in $A^{(2)}$ such that for every element $\{i, j\}$ in $X^{(2)}$,

$$\alpha'\{i, j\} = \{\alpha i, \alpha j\}. \qquad (4.1.4)$$

Thus the degree of $A^{(2)}$ is $\binom{p}{2}$ and $A \cong A^{(2)}$ unless $A = S_2$. To clarify this definition we introduce the line-group of a graph, a concept which will also be rather useful later. Let G be a graph with point set $V(G)$ and line set $X(G)$. Each permutation α in $\Gamma(G)$ induces a permutation α' acting on $X(G)$ in the following way. If u and v are adjacent in G so that $\{u, v\}$ is a line of G, then

$$\alpha'\{u, v\} = \{\alpha u, \alpha v\}. \qquad (4.1.5)$$

This collection of permutations of $X(G)$ constitutes a group, denoted $\Gamma_1(G)$, called the *line-group of* G. For example, the line-group of the complete graph of order p and the pair group of the symmetric group of degree p are identical,

$$\Gamma_1(K_p) = S_p^{(2)}. \qquad (4.1.6)$$

Now by (4.1.3) and (4.1.4), two functions f and h represent the same graph if and only if there is a permutation α of X such that for all z in $X^{(2)}$

$$f(z) = h(\alpha'z). \tag{4.1.7}$$

But this condition is precisely the requirement that f and h are in the same orbit of the power group $E_2^{S_p^{(2)}}$. Consequently the orbits of this power group correspond to the different unlabeled graphs of order p, that is, the isomorphism classes of labeled graphs. On applying PET to count these orbits by weight, Pólya's formula for counting graphs takes the following form.

Theorem   The polynomial $g_p(x)$ which enumerates graphs of order p by number of lines is given by

$$g_p(x) = Z(S_p^{(2)}, 1 + x), \tag{4.1.8}$$

where

$$Z(S_p^{(2)}) = \frac{1}{p!} \sum_{(j)} \frac{p!}{\prod_k k^{j_k} j_k!} \prod_k s_{2k+1}^{k j_{2k+1}} \prod_k (s_k s_{2k}^{k-1})^{j_{2k}} s_k^{k\binom{j_k}{2}} \prod_{r<t} s_{[r,t]}^{(r,t)j_r j_t}. \tag{4.1.9}$$

Proof   Define the weight function w on $Y = \{0, 1\}$ by setting $w(0) = 0$ and $w(1) = 1$. Then the weight of a function from $X^{(2)}$ to Y and the weight of the orbit to which it belongs are defined in the usual way, as in (2.4.4). Thus the weight of an orbit of $E_2^{S_p^{(2)}}$ is simply the number of lines in the corresponding graph. Then (4.1.8) follows from PET.

Now we shall derive formula (4.1.9) for $Z(S_p^{(2)})$ but we illustrate first with $p = 4$. In the cycle index

$$Z(S_4) = \tfrac{1}{24}(s_1^4 + 6s_1^2 s_2 + 8s_1 s_3 + 3s_2^2 + 6s_4) \tag{4.1.10}$$

each term can be represented by one of its contributing permutations as in Table 4.4.1.

Each of the five types of permutations in S_4 induces a permutation in the pair group $S_4^{(2)}$. The next table shows the terms in $Z(S_4^{(2)})$ induced by each term of $Z(S_4)$.

One of these terms is now illustrated. In Figure 4.1.1, the 2-subsets are indicated for brevity by juxtaposition: $\{1, 2\} = 12$, and so on. Collecting terms, we find that

$$Z(S_4^{(2)}) = \tfrac{1}{24}(s_1^6 + 9s_1^2 s_2^2 + 8s_3^2 + 6s_2 s_4). \tag{4.1.11}$$

On substituting $1 + x$ in this cycle index, one has as in (4.1.2),

$$Z(S_4^{(2)}, 1 + x) = g_4(x) = 1 + x + 2x^2 + 3x^3 + 2x^4 + x^5 + x^6. \tag{4.1.12}$$

TABLE 4.1.1

Term of $Z(S_4)$	Permutation of S_4	Diagram of this permutation
s_1^4	(1)(2)(3)(4)	
$s_1^2 s_2$	(1)(2)(34)	
$s_1 s_3$	(1)(234)	
s_2^2	(12)(34)	
s_4	(1234)	

TABLE 4.1.2

Term of $Z(S_4)$	Induced term of $Z(S_4^{(2)})$
s_1^4	s_1^6
$s_1^2 s_2$	$s_1^2 s_2^2$
$s_1 s_3$	s_3^2
s_2^2	$s_1^2 s_2^2$
s_4	$s_2 s_4$

Figure 4.1.1

The permutation $(1)(2)(34)$ and its induced permutation in $S_4^{(2)}$.

We indicate the correspondence between terms of the cycle indexes of S_p and $S_p^{(2)}$ for $p = 4$ by writing

$$s_1^4 \to s_1^6, \qquad s_1^2 s_2 \to s_1^2 s_2^2,$$

and so on. To obtain an expression for $Z(S_p^{(2)})$, we must find the missing right-hand member of

$$s_1^{j_1} s_2^{j_2} \dots s_p^{j_p} \to \; ? \tag{4.1.13}$$

Let α be a permutation in S_p whose contribution to $Z(S_p)$ is $\prod s_k^{j_k}$. There are two separate contributions made by α' to the corresponding term in $Z(S_p^{(2)})$. The first comes from pairs of elements of $X = \{1, \dots, p\}$, both in a common cycle of α; the second, from pairs of elements of X, one in each of two different cycles of α.

We now determine the first of these contributions. Let $z_k = (12 \dots k)$ be a cycle of length k in α. Figure 4.1.2 shows the permutation in the pair group induced by z_k for $k = 2$ through 6. Observe that if k is odd, z_k induces $(k - 1)/2$ cycles of the same length,

$$s_k \to s_k^{(k-1)/2}.$$

On the other hand, when k is even, we find

$$s_k \to s_{k/2} s_k^{(k-2)/2}.$$

Thus since there are j_k cycles of length k in α, the pairs of elements lying in common cycles contribute

$$s_k^{j_k} \to s_k^{j_k(k-1)/2} \tag{4.1.14}$$

for k odd and

$$s_k^{j_k} \to \left(s_{k/2} s_k^{(k-2)/2}\right)^{j_k} \tag{4.1.15}$$

for k even.

To calculate the second contribution, consider two cycles z_r and z_t in α. As usual, let $[r, t]$ and (r, t) denote the l.c.m. and g.c.d. respectively. Then

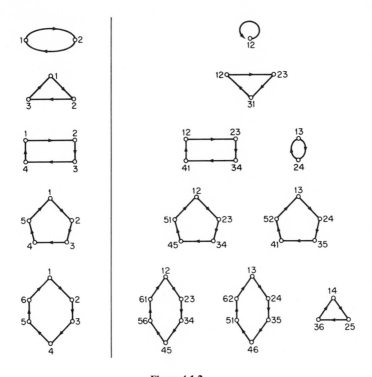

Figure 4.1.2

Cycles in S_p and the induced permutation in $S_p^{(2)}$.

the cycles z_r and z_t induce on the pairs of elements, one from each cycle, exactly (r, t) cycles of length $[r, t]$. In particular when $r = t = k$, they contribute k cycles of length k. Thus when $r \neq t$, we have

$$s_r^{j_r} s_t^{j_t} \rightarrow s_{[r,t]}^{(r,t)j_r j_t}, \qquad (4.1.16)$$

and when $r = t = k$,

$$s_k^{j_k} \rightarrow s_k^{k\binom{j_k}{2}}. \qquad (4.1.17)$$

Now on multiplying the right sides of (4.1.14)–(4.1.17) over all applicable cases, the missing right-hand member of (4.1.13) is obtained and (4.1.9) is verified.   //

Formulas (4.1.8) and (4.1.9) have been used (see Riordan [R15, p. 146]) to determine the polynomials $g_p(x)$ for $p \leq 9$. The coefficients for $p \leq 10$ are given in Appendix I. The cycle indexes $Z(S_p^{(2)})$ with $p \leq 10$ are also in

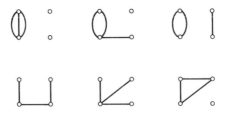

Figure 4.1.3

The six multigraphs of order 4 with three lines.

Appendix I and we shall see that they play an important role in the enumeration of several kinds of graphs.

In a *multigraph* more than one line can join two points (see Figure 4.1.3). Let $m_p(x)$ be the generating function for multigraphs so that the coefficient of x^q is the number of multigraphs with p points and q lines. The following theorem of [H4] was obtained by modifying the figure counting series for graphs.

Theorem The generating function $m_p(x)$ for multigraphs of order p is given by

$$m_p(x) = Z(S_p^{(2)}, 1/(1 - x)). \tag{4.1.18}$$

With $p = 4$, we find

$$m_4(x) = 1 + x + 3x^2 + 6x^3 + 11x^4 + 18x^5 + 32x^6$$
$$+ 48x^7 + 75x^8 + 111x^9 + 160x^{10} \cdots . \tag{4.1.19}$$

The coefficient of x^3 is verified by Figure 4.1.3.

We can now describe the enumeration of locally restricted graphs following Parthasarathy [P7] and [HP3]. Essentially this amounts to a further refinement of the cycle index of the pair group of S_p. The *partition* of a graph is the sequence of degrees of its points, usually written in descending order. A *locally restricted* graph is a graph with a given partition. The generating function that enumerates locally restricted graphs with p points is a polynomial $N(x_1, x_2, \ldots, x_p)$ such that each term, $x_1^{e_1} x_2^{e_2} \ldots x_p^{e_p}$ satisfies

$$e_1 \geq e_2 \geq \cdots \geq e_p.$$

The coefficient of such a term is the number of graphs with partition $e_1, e_2, \ldots, e_p$. For example, the term in $N(x_1, \ldots, x_5)$ that corresponds to the graph in Figure 4.1.4 is $x_1^4 x_2^3 x_3^3 x_4^2 x_5^2$ since its degree sequence is 4, 3, 3, 2, 2.

Figure 4.1.4

A graph with degree sequence 4, 3, 3, 2, 2.

In obtaining a formula for the generating function, we use the natural setting provided by the power group as applied to this situation. With $Y = \{0, 1\}$ as above, consider the function W from $Y^{X^{(2)}}$ into the ring of polynomials in the variables x_i defined by

$$W(f) = \prod_{\{i,j\} \in X^{(2)}} (x_i x_j)^{f(\{i,j\})}. \tag{4.1.20}$$

Then in the graph of f, the degree of the point i is given by the exponent of x_i in $W(f)$. It is convenient to define the linear operator θ acting on $W(f)$ by specifying that for any monomial in the x_i, θ reorders the exponents in nonincreasing order while stating the variables in increasing order. For example,

$$\theta(x_3^4 x_1^2 x_2^3 x_4) = x_1^4 x_2^3 x_3^2 x_4.$$

An application of the weighted version of Burnside's Lemma (2.3.10) gives the following result, which was obtained by Parthasarathy in somewhat different form.

Theorem The generating function that enumerates locally restricted graphs is

$$N(x_1, x_2, \ldots, x_p) = \frac{1}{p!} \sum_{\gamma \in E_2^{S_p^{(2)}}} \theta\left(\sum_{f = \gamma f} W(f)\right), \tag{4.1.21}$$

where

$$\sum_{f = \gamma f} W(f) = \prod_{z_r, z_s} \left(1 + \prod_{i \in z_r} x_i^{s/(r,s)} \prod_{j \in z_s} x_j^{r/(r,s)}\right)^{(r,s)}$$

$$\times \prod_{\substack{z_r \\ r \text{ even}}} \left(1 + \prod_{i \in z_r} x_i\right)\left(1 + \prod_{i \in z_r} x_i^2\right)^{(r-2)/2}$$

$$\times \prod_{\substack{z_r \\ r \text{ odd}}} \left(1 + \prod_{i \in z_r} x_i^2\right)^{(r-1)/2} \tag{4.1.22}$$

in which the first product is over all distinct pairs of cycles r, s in α and the others are over all cycles of the permutation in S_p which corresponds to γ.

Now by applying this theorem, we have

$$
\begin{aligned}
N(x_1, x_2, x_3) &= (1/3!)\{\theta(1 + x_1x_2)(1 + x_2x_3)(1 + x_1x_3) \\
&\quad + 3\theta(1 + x_1x_2x_3^2)(1 + x_1x_2) + 2\theta(1 + x_1^2x_2^2x_3^2)\} \\
&= (1/3!)\{(1 + 3x_1x_2 + 3x_1^2x_2x_3 + x_1^2x_2^2x_3^2) \\
&\quad + 3(1 + x_1x_2 + x_1^2x_2x_3 + x_1^2x_2^2x_3^2) + 2(1 + x_1^2x_2^2x_3^2)\} \\
&= 1 + x_1x_2 + x_1^2x_2x_3 + x_1^2x_2^2x_3^2.
\end{aligned}
$$

One must realize that this method only gives a formal solution and does not conveniently yield exact numbers or orders of magnitude.

4.2   CONNECTED GRAPHS

As in the case of labeled graphs, we shall see that the generating functions for graphs and connected graphs are closely related by the exponential function. Furthermore, connected graphs with specified properties can often be enumerated in a similar manner in terms of the total number of such graphs.

Let $g(x)$ be the generating function for graphs so that

$$
g(x) = \sum_{p=1}^{\infty} g_p x^p, \tag{4.2.1}
$$

where g_p is the number of graphs of order p, and let $c(x)$ be the corresponding generating function for connected graphs,

$$
c(x) = \sum_{p=1}^{\infty} c_p x^p. \tag{4.2.2}
$$

The theorem of Riddell [R14] which relates these two power series can now be stated.

Theorem   The generating functions $g(x)$ and $c(x)$ for graphs and connected graphs satisfy

$$
1 + g(x) = \exp \sum_{k=1}^{\infty} c(x^k)/k. \tag{4.2.3}
$$

Proof It follows from PET that $Z(S_n, c(x))$ counts graphs with exactly n components. Hence on summing this series over n,

$$1 + g(x) = Z(S_\infty, c(x)). \tag{4.2.4}$$

Then (4.2.3) follows from (4.2.4) and the identity (3.1.1).　//

Implicit in (4.2.3) is an effective method developed by Cadogan [C1] for computing the number of connected graphs of order p. First we set

$$\sum_{p=1}^{\infty} a_p x^p = \log(1 + g(x)). \tag{4.2.5}$$

Then from (1.2.8), it follows that

$$pa_p = pg_p - \sum_{k=1}^{p-1} ka_k g_{p-k}. \tag{4.2.6}$$

It is from this equation and the values of g_p that the integers pa_p are first calculated in Table 4.2.1 for $p \leq 9$. Using them, we next see how the values of c_p are obtained.

TABLE 4.2.1

THE NUMBER OF CONNECTED GRAPHS

p	g_p	pa_p	c_p
1	1	1	1
2	2	3	1
3	4	7	2
4	11	27	6
5	34	106	21
6	156	681	112
7	1 044	5 972	853
8	12 346	88 963	11 117
9	274 668	2 349 727	261 080

Since

$$\sum_{p=1}^{\infty} a_p x^p = \sum_{k=1}^{\infty} c(x^k)/k. \tag{4.2.7}$$

it follows by equating coefficients that

$$pa_p = \sum_{d \mid p} dc_d. \tag{4.2.8}$$

On inverting (4.2.8) using the möbius function $\mu(d)$, the numbers can be expressed in terms of the a_p,

$$c_p = \sum_{d \mid p} \frac{\mu(d)}{d} a_{p/d}. \qquad (4.2.9)$$

This formula of Cadogan was used to calculate the values of c_p in Table 4.2.1 with $p \leq 9$. Whenever *any* two generating functions satisfy the relation (4.2.3) of the theorem, the coefficients are related by (4.2.6) and (4.2.9). It is often the case that the generating functions for graphs with a specified property and for such connected graphs satisfy (4.2.3). Hence the connected graphs under consideration can be enumerated using (4.2.6) and (4.2.9). For example, if $w(x)$ is the generating function for the even graphs defined in Chapter 1 and $u(x)$ counts connected, even graphs, then

$$1 + w(x) = \exp \sum_{k=1}^{\infty} u(x^k)/k. \qquad (4.2.10)$$

Hence the coefficients of $u(x)$ can be computed from those of $w(x)$ by means of the relations corresponding to (4.2.6) and (4.2.9). The connected, even graphs are precisely the eulerian graphs, and we shall provide the details of the computations involved in the last section of this chapter.

Cadogan [C1] extended his method to include lines as well as points as an enumeration parameter. We now sketch the details of this process. Let $g_{p,q}$ and $c_{p,q}$ be the number of (p, q) graphs and connected graphs respectively, and set

$$g(x, y) = \sum g_{p,q} x^p y^q \qquad (4.2.11)$$

$$c(x, y) = \sum c_{p,q} x^p y^q. \qquad (4.2.12)$$

Then it follows from the two variable version of PET that

$$1 + g(x, y) = \exp \sum_{k=1}^{\infty} c(x^k, y^k)/k. \qquad (4.2.13)$$

For each $p \geq 1$, let $b_p(y)$ be the polynomial in y defined by

$$\sum_{p=1}^{\infty} b_p(y) x^p = \log(1 + g(x, y)). \qquad (4.2.14)$$

Then it can be shown as in (4.2.6) that

$$p b_p(y) = p g_p(y) - \sum_{k=1}^{p-1} k b_k(y) g_{p-k}(y). \qquad (4.2.15)$$

The coefficients of the $b_p(y)$ can be computed from the coefficients of the $g_p(y)$ using (4.2.15). For convenience, we set

$$b_p(y) = \sum b_{p,q} y^q. \tag{4.2.16}$$

We also have

$$\sum b_p(y)x^p = \sum_{k=1}^{\infty} c(x^k, y^k)/k, \tag{4.2.17}$$

and on equating coefficients of $x^p y^q$ and using möbius inversion we have

$$c_{p,q} = \sum_{r|(p,q)} b_{p/r,q/r}\mu(r)/r. \tag{4.2.18}$$

The values of $c_{p,q}$ in Table 4.2.2 were computed by Cadogan using (4.2.18).

Clearly this approach also can be used to determine the number of connected (p, q) graphs with specified properties.

TABLE 4.2.2

THE NUMBER OF CONNECTED (p, q) GRAPHS

p \ q	0	1	2	3	4	5	6	7	8	9	10	11	12	13
1	1													
2		1												
3			1	1										
4				2	2	1	1							
5					3	5	5	4	2	1	1			
6						6	13	19	22	20	14	9	5	2
7							11	33	67	107	132	138	126	95
8								23	89	236	486	814	1 169	1 454

4.3 BICOLORED GRAPHS

We have seen in Chapter 1 that the points of a k-colored graph have been partitioned into k sets so that adjacent points are always in different sets. The points of each set are then considered to have the same color. In this section we concentrate on *bicolored graphs* for which $k = 2$.

Substituting (4.1.6) into (4.1.8), we have

$$g_p(x) = Z(\Gamma_1(K_p), 1 + x). \tag{4.3.1}$$

It is now fruitful to generalize this equation by replacing K_p by an arbitrary graph G of order p, as in [H11]. It follows from Corollary (2.5.1) of the PET, which interprets $Z(A, 1 + x)$, that $Z(\Gamma_1(G), \ 1 + x)$ enumerates $\Gamma_1(G)$-equivalence classes of sets of lines of G. These equivalence classes correspond precisely to spanning subgraphs of G, two of which are in the same class whenever there is an automorphism of G that sends one to the other. If two subgraphs are not in the same class, they are called *dissimilar*.

Theorem The number of dissimilar spanning subgraphs of G with q lines is the coefficient of x^q in $Z(\Gamma_1(G), 1 + x)$.   (4.3.2)

To illustrate we shall consider the (4, 5) graph G shown in Figure 4.3.1. Routine calculation shows that

$$Z(\Gamma_1(G)) = \tfrac{1}{4}(s_1^5 + 3s_1 s_2^2), \tag{4.3.3}$$

and hence

$$Z(\Gamma_1(G), 1 + x) = 1 + 2x + 4x^2 + 4x^3 + 2x^4 + x^5. \tag{4.3.4}$$

These coefficients are illustrated in the next figure. We have used dashed lines to indicate the missing lines, to emphasize the equivalence with respect to the group of G. Note that (4.3.4) also gives the number of 2-colorings of the lines G (solid and dashed colors). We shall find this interpretation of $Z(\Gamma_1(G), \ 1 + x)$ very useful, particularly in counting bicolored graphs, which follows next.

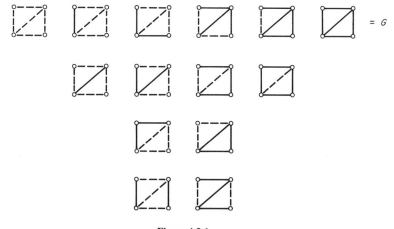

Figure 4.3.1

The spanning subgraphs of a random graph.

The *complete bipartite graph* $K_{m,n}$ has $m + n$ points of which m have one color and n have the other color with two points adjacent if and only if they have different colors. Since bicolored graphs with m points of one color and n of the other correspond precisely to spanning subgraphs of $K_{m,n}$, it follows immediately from the Theorem (4.3.2) that the polynomial $b_{m,n}(x)$ which counts these bicolored graphs satisfies

$$b_{m,n}(x) = Z(\Gamma_1(K_{m,n}), 1 + x). \tag{4.3.5}$$

For example,

$$b_{3,2}(x) = 1 + x + 3x^2 + 3x^3 + 3x^4 + x^5 + x^6, \tag{4.3.6}$$

and the next figure verifies these coefficients. Note that the coefficients of these polynomials are always end-symmetric.

We now determine the cycle index of the line-group of $K_{m,n}$, following [H7]. It is much simpler to handle first the case $m \neq n$. Suppose X is the set of m points of one color in $K_{m,n}$ and Y is the set of n points of the other color. Then the ordered pairs (x, y) in the cartesian product $X \times Y$ correspond precisely to the lines of $K_{m,n}$. Thus the permutations in $\Gamma_1(K_{m,n})$ consist of permutations of the pairs (x, y) induced by permutations of X and of Y. This suggests the following binary operation on permutation groups. Let A and B be permutation groups with object sets X and Y respectively. The *cartesian product of A and B*, denoted $A \times B$, is a permutation group with object set $X \times Y$. Its permutations consist of all ordered pairs (α, β) of permutations α in A and β in B. The image of each element (x, y) of $X \times Y$ determined by (α, β) is

$$(\alpha, \beta)(x, y) = (\alpha x, \beta y). \tag{4.3.7}$$

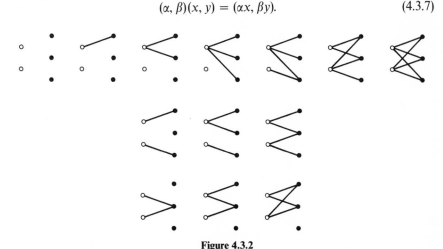

Figure 4.3.2

The bicolored graphs with three points of one color and two of the other color.

Therefore for $m \neq n$,

$$\Gamma_1(K_{m,n}) = S_m \times S_n, \tag{4.3.8}$$

and we have the theorem of [H7] which counts these bicolored graphs by substituting (4.3.8) into (4.3.5).

Theorem The polynomial $b_{m,n}(x)$ that enumerates bicolored graphs with $m \neq n$ is given by

$$b_{m,n}(x) = Z(S_m \times S_n, 1 + x) \tag{4.3.9}$$

where

$$Z(S_m \times S_n) = \frac{1}{m!n!} \sum_{(\alpha,\beta)} \prod_{r,t=1}^{m,n} s_{[r,t]}^{(r,t)j_r(\alpha)j_t(\beta)}. \tag{4.3.10}$$

Actually, we have already calculated the cycle index of $S_m \times S_n$ within the formula (4.1.9) for $Z(S_p^{(2)})$. In particular, (4.1.16) gives precisely the term under the sum and product signs in (4.3.10).

To illustrate the theorem, we shall use formulas (4.3.9) and (4.3.10) to derive $b_{3,2}(x)$ already given in (4.3.6). First we shall find from (4.3.10) the cycle index of the cartesian product $S_3 \times S_2$. There are twelve permutations in this group, but the cycle structure of any one of them, say (α, β), depends only on the cycle structures of α and β. The cycle structures of these individual permutations are obtained from formula (2.2.5) which expresses $Z(S_n)$ in terms of partitions of n. There are only three partitions of 3 and two of 2; so we have just six different kinds of pairs of permutations to consider. For example, if the permutation α in S_3 contributes the monomial $s_1 s_2$ to $Z(S_3)$ and β in S_2 is the transposition, then it follows from (4.3.10) that (α, β) contributes s_2^3 to $Z(S_3 \times S_2)$. Since there are three permutations with the same cycle structure as α, their total contribution to $Z(S_3 \times S_2)$ when paired with β is $3s_2^3$. Table 4.3.1 contains the contributions for all pairs in $S_2 \times S_3$.

TABLE 4.3.1

THE TERMS OF $Z(S_3 \times S_2)$

Permutation	α	β	(α, β)
Cycle index term	s_1^3	s_1^2	s_1^6
	$s_1 s_2$	s_1^2	$s_1^2 s_2^2$
	s_3	s_1^2	s_3^2
	s_1^3	s_2	s_2^3
	$s_1 s_2$	s_2	s_2^3
	s_3	s_2	s_6

Hence the cycle index formula is

$$Z(S_3 \times S_2) = \tfrac{1}{12}(s_1^6 + 3s_1^2 s_2^2 + 2s_3^2 + 4s_2^3 + 2s_6). \qquad (4.3.11)$$

On substituting $1 + x^k$ for s_k in (4.3.11); the polynomial $b_{3,2}(x)$ of (4.3.6) is obtained.

We now turn to the enumeration of bicolored graphs for the more subtle case $m = n$. The corresponding counting polynomial is denoted by $b_n(x)$ instead of $b_{n,n}(x)$. For example, it is not hard to verify that

$$b_3(x) = 1 + x + 2x^2 + 4x^3 + 5x^4 + 5x^5 + 4x^6 + 2x^7 + x^8 + x^9. \qquad (4.3.12)$$

Figure 4.3.3 verifies that the coefficient of x^4, and hence that of x^5, is 5. Note well that we do not obtain additional isomorphism classes of these bigraphs by interchanging the colors.

As before we must find a formula for the cycle index of the line-group of $K_{n,n}$. For this purpose it is convenient to define additional operations on permutation groups.

Let A and B be permutation groups with object sets $X = \{x_1, \ldots, x_d\}$ and $Y = \{y_1, \ldots, y_e\}$. The *composition of A with B* is denoted by $A[B]$ and has $X \times Y$ as its object set. It was defined by Pólya [P8] and others and is sometimes called the "wreath product." For each permutation α in A and each sequence $\beta_1, \ldots, \beta_d$ of d permutations in B, there is a permutation in $A[B]$, denoted $[\alpha; \beta_1, \ldots, \beta_d]$, such that for every ordered pair (x_i, y_j) in $X \times Y$,

$$[\alpha; \beta_1, \ldots, \beta_d](x_i, y_j) = (\alpha x_i, \beta_i y_j). \qquad (4.3.13)$$

The reason for calling this group the composition of A and B is that its cycle index is the composition of $Z(A)$ with $Z(B)$ in the manner illustrated by the following example. Take $A = C_3$ and $B = S_2$. Using different letters for the variables in $Z(A)$ and $Z(B)$, we have

$$Z(A) = \tfrac{1}{3}(s_1^3 + 2s_3), \qquad Z(B) = \tfrac{1}{2}(t_1^2 + t_2)$$

$$Z(A[B]) = \tfrac{1}{3}([\tfrac{1}{2}(t_1^2 + t_2)]^3 + 2[\tfrac{1}{2}(t_3^2 + t_6)]).$$

The sense in which the cycle index of the composition of two groups is the composition of their cycle indexes should now be apparent. Pólya proved

Figure 4.3.3

The five (6,4) bicolored graphs with three points of each color.

that $Z(A[B])$ is that polynomial obtained from $Z(A)$ by replacing each variable s_k by $Z(B; s_k, s_{2k}, s_{3k}, \ldots)$. Thus $Z(B)$ has been substituted into the cycle index of A to obtain the cycle index of their composition. We often denote this substitution by

$$Z(A[B]) = Z(A)[Z(B)]. \qquad (4.3.14)$$

Furthermore, for any two permutation groups A and B and for any power series $f(x)$, we have

$$Z(A[B], f(x)) = Z(A, Z(B, f(x))). \qquad (4.3.15)$$

The most typical example of the composition of two permutation groups is obtained from the graph that consists of m disjoint copies of K_n. Its group is identically $S_m[S_n]$. As noted in (2.1.1), a graph G and its complement $\bar{G}$ have the same group. Hence to recognize the group of $K_{n,n}$, we use the fact that its complement consists of two copies of K_n and get

$$\Gamma(K_{n,n}) = S_2[S_n]. \qquad (4.3.16)$$

Since $\Gamma_1(K_{n,n})$ is induced by $\Gamma(K_{n,n})$ we can now use (4.3.16) to determine $Z(\Gamma_1(K_{n,n}))$. The permutations in $\Gamma(K_{n,n})$ that fix the two n-sets of different colors correspond to the terms in $Z(S_2[S_n])$ obtained by substituting $Z(S_n)$ for s_1 in $Z(S_2)$. Their contribution to $Z(\Gamma_1(K_{n,n}))$ is precisely $Z(S_n \times S_n)$. The permutations in $\Gamma(K_{n,n})$ which interchange the two n-sets contribute $Z(S_n; s_2, s_4, s_6, \ldots)$ to $Z(\Gamma(K_{n,n}))$. Thus they consist of cycles of even length which always permute a point of one color to a point of the other color. Their contribution to $Z(\Gamma_1(K_{n,n}))$ is shown in [H7] to be

$$Z'_n = \frac{1}{n!} \sum_{(j)} \frac{n!}{\prod k^{j_k} j_k!} \prod_{k \text{ odd}} s_k^{j_k} \prod_k s_{2k}^{k\binom{j_k}{2} + [k/2]j_k} \prod_{r<t} s_{2[r,t]}^{(r,t)j_r j_t}. \qquad (4.3.17)$$

This formula can be verified by considering cycles of points with alternating colors and by observing how these cycles permute all the lines which join points of different colors.

We have now found that

$$Z(\Gamma_1(K_{n,n})) = \tfrac{1}{2}(Z(S_n \times S_n) + Z'_n), \qquad (4.3.18)$$

and can write a formula for this cycle index using (4.3.10) and (4.3.17). However, there is another useful operation on permutation groups [H9], a special case of which provides a group identical to $\Gamma_1(K_{n,n})$. Let A and B be permutation groups with object sets $X = \{x_1, x_2, \ldots, x_d\}$ and $Y = \{y_1, y_2, \ldots, y_e\}$ as above, but now we insist that $e > 1$. The *exponentiation* $[B]^A$ *of* A *with* B is that permutation group with object set Y^X whose permutations are constructed as follows. Each permutation α in A and each sequence $\beta_1, \ldots, \beta_d$ of permutations in B determine just one permutation $(\alpha; \beta_1, \ldots, \beta_d)$ in $[B]^A$, which takes the function f into the function f^* defined for all $x_i \in X$ by

$$f^*(x_i) = \beta_i f(\alpha x_i). \tag{4.3.19}$$

It can easily be shown that the distinct selections of $\alpha, \beta_1, \ldots, \beta_d$ lead to distinct permutations of Y^X and that these permutations form a group. Furthermore, when we take $A = S_2$ and $B = S_n$, the functions Y^X correspond precisely to the lines of $K_{n,n}$ and in fact

$$\Gamma_1(K_{n,n}) = [S_n]^{S_2}. \tag{4.3.20}$$

Then the counting theorem of [H7] takes the following form.

Theorem The polynomial $b_n(x)$ that enumerates bicolored graphs with n points of each color is given by

$$b_n(x) = Z([S_n]^{S_2}, 1 + x) \tag{4.3.21}$$

where

$$Z([S_n]^{S_2}) = \tfrac{1}{2}(Z(S_n \times S_n) + Z'_n). \tag{4.3.22}$$

As mentioned above, this theorem counts bigraphs with n points of each color in which the colors are interchangeable. The simpler case, with fixed colors, was already handled in (4.3.9) by taking $m = n$, even though the theorem says not to.

We now illustrate the theorem by providing the formulas required to determine $b_3(x)$. The cycle index of $S_3 \times S_3$ is obtained by applying (4.3.10) to get

$$Z(S_3 \times S_3) = \tfrac{1}{36}(s_1^9 + 6s_1^3 s_2^3 + 8s_3^3 + 9s_1 s_2^4 + 12s_3 s_6). \tag{4.3.23}$$

From formula (4.3.17) it follows that

$$Z'_3 = \tfrac{1}{6}(s_1^3 s_2^3 + 3s_1 s_4^2 + 2s_3 s_6). \tag{4.3.24}$$

Combining these we have

$$Z([S_3]^{S_2}) = \tfrac{1}{72}(s_1^9 + 12s_1^3 s_2^3 + 8s_3^3 + 9s_1 s_2^4 + 18s_1 s_4^2 + 24s_3 s_6). \tag{4.3.25}$$

Now on substituting $1 + x^k$ for s_k in this expression, formula (4.3.12) for $b_3(x)$ is obtained.

A *bicolorable graph* is one which can be bicolored. We only indicate how such graphs have been counted. As observed in [HP15], the number of connected bicolored graphs of order p equals the number of connected bicolorable graphs of order p. Having enumerated bicolored graphs, we can count the connected ones using the method of (4.2.3), thus determining the number of connected bicolorable graphs as well.

4.4 ROOTED GRAPHS

Rooted labeled graphs were counted effortlessly within the proof of (1.2.1). The enumeration of rooted (unlabeled) graphs is a bit less obvious [H4]. It involves the formation of a pair group slightly altered from that of the symmetric group. This modification generalizes readily to graphs rooted at an arbitrary induced subgraph [HP1], which may be a graph, digraph, or multigraph.

There is just one theorem in this section, namely, a formula for counting graphs G of order p rooted at a specified induced subgraph H, where G has q lines in addition to those of H. As illustrative corollaries, we will give formulas for rooted graphs, line-rooted graphs, triangle-rooted graphs, graphs rooted at an oriented line, and graphs rooted at a cyclic triple. For example, all the (6, 5) graphs rooted at a triangle $H = K_3$ are displayed in the next figure where dashed lines are used for the triangle. Thus by the above convention, $p = 6$ and $q = 2$ since there are two additional lines.

For any set S of points of the graph G, the *induced subgraph* $\langle S \rangle$ is the maximal subgraph of G with point set S. Thus $\langle S \rangle$ contains all lines of G joining two points in S. We seek to determine the number of graphs G of order p rooted at an induced subgraph H of order $n < p$. This means that no line of G joins two nonadjacent points of H.

We shall denote the permutation group required for the application of PET via (2.5.1) by $\Gamma(H) \circ S_{p-n}$, defined as follows: We set $G = H \cup K_{p-n}$ and observe that the product $\Gamma(H)S_{p-n}$ is the subgroup of $\Gamma(G)$ which fixes H. We require the restriction of the pair group $(\Gamma(H)S_{p-n})^{(2)}$ to those unordered pairs $\{u, v\}$ of points of G not both in H. The resulting permutation group is designated by $\Gamma(H) \circ S_{p-n}$.

If the contribution of a permutation to $Z(S_p)$ is $\prod s_k^{j_k}$, then the corresponding contribution to $Z(S_p^{(2)})$ is denoted by $(\prod s_k^{j_k})^{(2)}$. Now that the necessary symbols are defined, the general result [HP1] can be stated.

Theorem The counting polynomial $h_p(x)$ that enumerates graphs of order p rooted at an induced subgraph H of order $n < p$ is

$$h_p(x) = Z(\Gamma(H) \circ S_{p-n}, 1 + x) \qquad (4.4.1)$$

where

$$Z(\Gamma(H) \circ S_{p-n}) = \frac{1}{|\Gamma(H)|(p-n)!} \sum_{(\alpha,\beta)} \prod_{r,t=1}^{n,p-n} s_{[r,t]}^{(r,t)j_r(\alpha)j_t(\beta)} (\prod_k s_k^{j_k(\beta)})^{(2)} \tag{4.4.2}$$

and the sum is over all pairs with α in $\Gamma(H)$ and β in S_{p-n}.

Proof The permutations in $\Gamma(H) \circ S_{p-n}$ are induced by the pairs (α, β) of permutations with α in $\Gamma(H)$ and β in S_{p-n}. The contribution to $Z(\Gamma(H) \circ S_{p-n})$ of any pair can be considered as a product of two terms. In the first term we just observe that (α, β) induces a permutation on the lines between H and K_{p-n} exactly as in the cartesian product (4.3.10). The second term is $(\prod_k s_k^{j_k(\beta)})^{(2)}$ which describes the structure of the permutation β induces on the pairs of points in K_{p-n}. //

We can now squeeze some corollaries out of the theorem by choosing special kinds of subgroups for H. For our first application, we consider graphs of order p rooted at a point. Thus $H = K_1$ so $\Gamma(H) = S_1$, and clearly $\Gamma(H) \circ S_{p-n} = (S_1 S_{p-1})^{(2)}$. Substituting into (4.4.1), we get the generating function $G_p(x)$ for rooted graphs

$$G_p(x) = Z((S_1 S_{p-1})^{(2)}, 1 + x), \tag{4.4.3}$$

which is, of course, the same result as in [H4].

For only three points, we quickly have

$$G_3(x) = Z((S_1 S_2)^{(2)}, 1 + x) = \tfrac{1}{2}[(1 + x)^3 + (1 + x)(1 + x^2)]$$
$$= 1 + 2x + 2x^2 + x^3.$$

To illustrate with $p = 4$, we observe that

$$Z(S_1 S_3) = (1/3!)(s_1^4 + 3s_1^2 s_2 + 2s_1 s_3), \tag{4.4.4}$$

and making use of Table 4.1.2 yields

$$Z((S_1 S_3)^{(2)}) = (1/3!)(s_1^6 + 3s_1^2 s_2^2 + 2s_3^2). \tag{4.4.5}$$

On substituting $1 + x$ in this cycle index, the result is

$$G_4(x) = 1 + 2x + 4x^2 + 6x^3 + 4x^4 + 2x^5 + x^6, \tag{4.4.6}$$

which is easily verified by inspection of the graphs of order 4 in Figure 1.1.2. In general one can make use of the cycle index formulas in Appendix III, as far as it goes, to obtain $Z((S_1 S_{p-1})^{(2)})$.

In our next illustration we take $H = K_2$ so that $h_p(x)$ counts graphs rooted at a line:

$$h_p(x) = Z(S_2 \circ S_{p-2}, 1 + x). \tag{4.4.7}$$

The identity

$$Z(S_2 \circ S_{p-2}) = (\partial/\partial s_1)Z(S_p^{(2)}) \tag{4.4.8}$$

can be quickly verified. Hence we can use the cycle index formulas in Appendix III to obtain for $p = 5$:

$$Z(S_2 \circ S_3) = \tfrac{1}{12}(s_1^9 + 4s_1^3 s_2^3 + 2s_3^3 + 3s_1 s_2^4 + 2s_3 s_6). \tag{4.4.9}$$

On substituting $1 + x$ we find

$$h_5(x) = 1 + 2x + 6x^2 + 12x^3 + 16x^4 + 16x^5 + 12x^6$$
$$+ 6x^7 + 2x^8 + x^9. \tag{4.4.10}$$

The counting polynomial for graphs rooted at a triangle is also easily obtained. We give the details for such graphs with six points. Since $\Gamma(H) = S_3$, we use formula (4.4.2) to obtain

$$Z(S_3 \circ S_3) = \tfrac{1}{36}(s_1^{12} + 3s_1^4 s_2^4 + 6s_3^4 + 3s_1^6 s_2^3 + 9s_1^2 s_2^5$$
$$+ 6s_3^2 s_6 + 2s_3^3 s_1^3 + 6s_1 s_2 s_3 s_6). \tag{4.4.11}$$

From formula (4.4.1) we have

$$h_6(x) = 1 + 2x + 6x^2 + 15x^3 + 21x^4 + 38x^5 + 44x^6$$
$$+ 38x^7 + 21x^8 + 15x^9 + 6x^{10} + 2x^{11} + x^{12}.$$

The graphs enumerated by the coefficient of x^2 in this polynomial are shown in Figure 4.4.1, in which the lines of the root triangle are drawn using dashes.

Our theorem is also effective if H is a digraph. For example, if H is an oriented line, $\Gamma(H) = E_2$ and so

$$h_p(x) = Z(E_2 \circ S_{p-2}, 1 + x). \tag{4.4.12}$$

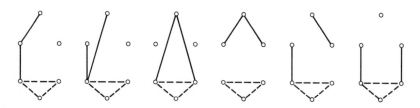

Figure 4.4.1

The (6,5) graphs rooted at a triangle.

An expression for this cycle index is obtained from (4.4.2). For each permutation β in S_{p-2}, let $\prod s_k^{j_k}$ be the contribution to $Z(S_{p-2})$. Then

$$Z(E_2 \circ S_{p-2}) = \frac{1}{(p-2)!} \sum_\beta (\prod s_k^{j_k})^2 (\prod s_k^{j_k})^{(2)} \qquad (4.4.13)$$

and the sum is over all β in S_{p-2}. To illustrate (4.4.13) with $p = 5$, recall that $Z(S_3^{(2)}) = Z(S_3)$,

$$Z(E_2 \circ S_3) = (1/3!)((s_1^3)^2 s_1^3 + 3(s_1 s_2)^2 s_1 s_2 + 2(s_3)^2 s_3). \qquad (4.4.14)$$

Then from (4.4.12)

$$h_5(x) = 1 + 3x + 9x^2 + 20x^3 + 27x^4 + 27x^5$$
$$+ 20x^6 + 9x^7 + 3x^8 + x^9. \qquad (4.4.15)$$

One can readily construct the graphs enumerated by these coefficients.

Finally we compute the counting polynomial for graphs with six points which are rooted at a cyclic triple. In this case

$$\Gamma(H) = \tfrac{1}{3}(s_1^3 + 2s_3). \qquad (4.4.16)$$

From (4.4.2) we obtain

$$Z(\Gamma(H) \circ S_3) = \tfrac{1}{18}(s_1^{12} + 3s_1^4 s_2^4 + 2s_1^3 s_3^3 + 6s_1 s_2 s_3 s_6 + 6s_3^4), \qquad (4.4.17)$$

and from (4.4.1) it follows that

$$h_6(x) = 1 + 2x + 6x^2 + 18x^3 + 34x^4 + 52x^5 + 62x^6$$
$$+ 52x^7 + 34x^8 + 18x^9 + 6x^{10} + 2x^{11} + x^{12}. \qquad (4.4.18)$$

The corresponding graphs with $q = 2$ are shown in Figure 4.4.2.

Of course many other graphs can be chosen for the root-subgraph H. In particular, whenever the graph or digraph H has the identity group E_n, the formula (4.2.2) reduces to

$$Z(E_n \circ S_{p-n}) = \frac{1}{(p-n)!} \sum_\beta (\prod s_k^{j_k})^n (\prod s_k^{j_k})^{(2)} \qquad (4.4.19)$$

where just as in (4.4.13), the sum is over all β in S_{p-n}.

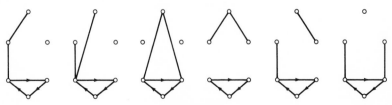

Figure 4.4.2

The (6,5) graphs rooted at a cyclic triple.

4.5 SUPERGRAPHS AND COLORED GRAPHS

Bicolored graphs were counted in Section 4.3 and graphs rooted at an induced subgraph were handled in the preceding section. Our present object is to extend both of these studies by counting graphs rooted at a subgraph which is not necessarily induced and then applying this to the enumeration of m-colored graphs. The results involve a kind of a neat combination of a rather pedestrian approach of [H5] with the elegant methodology of Robinson [R17].

If H is a subgraph of G, then G is a *supergraph* of H. Instead of the tedious phrase "the graphs G rooted at a subgraph H which is not necessarily induced," we prefer to say "the supergraphs of H," which provides a more effective mnemonic device.

We begin by obtaining a formula for the number of dissimilar p-point supergraphs of a graph H also of order p. We will see in the next equation that this involves the cycle index of the line-group of the complement of H. It follows from the method of the preceding section for enumerating graphs rooted at an induced subgraph H of order n, that the permutation group required for handling the supergraphs of H can be obtained by enlarging the object set of $\Gamma(H) \circ S_{p-n}$ to include the lines of $\overline{H}$. In particular, if H is a graph of order p, then the polynomial $r_p(x)$ that enumerates supergraphs rooted at H is given by

$$r_p(x) = Z(\Gamma_1(\overline{H}), 1 + x). \tag{4.5.1}$$

An interesting special case [H5] of (4.5.1) has H as the cycle C_p of order p. This serves to count the total number of dissimilar hamiltonian cycles appearing in all the (hamiltonian) graphs of order p. The line-group $\Gamma_1(\overline{C}_p)$ can be expressed in terms of wreath products (compositions). For example when $p = 2n + 1$, we have

$$\Gamma_1(\overline{C}_{2n+1}) = D_{2n+1}[E_{n-1}], \tag{4.5.2}$$

from which $Z(\Gamma_1(\overline{C}_{2n+1}))$ can be computed using formulas (2.2.11) and (4.3.14). When p is even, one can compute $Z(\Gamma_1(\overline{C}_{2n}))$ by multiplying each term of $Z(D_{2n}[E_{n-2}])$ by the appropriate term of $Z(D_n)$.

To illustrate, when $p = 5$, formula (4.5.2) becomes $\Gamma_1(\overline{C}_5) = D_5[E_1] = D_5$, and by (2.2.11)

$$Z(D_5) = \tfrac{1}{10}(s_1^5 + 4s_5 + 5s_1 s_2^2). \tag{4.5.3}$$

Then on applying (4.5.1), we have

$$r_5(x) = 1 + x + 2x^2 + 2x^3 + x^4 + x^5. \tag{4.5.4}$$

This polynomial is verified by the graphs of Figure 4.5.1 in which the root C_5 is the cycle with dashed lines.

Now we turn to a general method for enumerating supergraphs. Let H be any subgraph of order n of the complete graph K_p. Then the product $\Gamma(H)S_{p-n}$ is the group consisting of all permutations in $\Gamma(K_p) = S_p$ that fix H; that is, they permute the points of H among themselves. The group required for counting supergraphs of H is the restriction of the pair group $(\Gamma(H)S_{p-n})^{(2)}$ to the pairs of points which are not adjacent in H. We denote this restriction by $\Gamma_1(\bar{H}, K_p)$. Note that if H has p points, then $\Gamma_1(\bar{H}, K_p) = \Gamma_1(\bar{H})$. Furthermore if H is connected, then $\Gamma_1(\bar{H}, K_p)$ is just the line group of the complement of H in K_p.

Up to now, we have a theorem [H5] which determines the number of supergraphs of a given graph in principle, but no effective method for reckoning the required cycle index. Such an algorithm was provided by Robinson [R17]. As soon as this procedure is developed, we will have demonstrated the next theorem. Therefore it is particularly convenient at this juncture to present first the proof and then the theorem, contrary to the wise admonition of G. E. Uhlenbeck. In order to calculate the cycle index of $\Gamma_1(\bar{H}, K_p)$, we define the *point-line group* of a graph G, denoted $\Gamma_{0,1}(G)$, to be that permutation group induced by $\Gamma(G)$ which permutes both points *and* lines of G. When writing the cycle index of this group we shall distinguish the cycles of objects being permuted by using two sets of variables, s_k for points and t_k for lines. For example, if G is the (4, 5) graph $K_4 - x$, then

$$Z(\Gamma_{0,1}(K_4 - x)) = \tfrac{1}{4}(s_1^4 t_1^5 + 2s_1^2 s_2 t_1 t_2^2 + s_2^2 t_1 t_2^2) \qquad (4.5.5)$$

From (2.2.14) it follows that the cycle index of $\Gamma(H)S_{p-n}$ is given by

$$Z(\Gamma(H)S_{p-n}) = Z(\Gamma(H))Z(S_{p-n}). \qquad (4.5.6)$$

Similarly $Z(\Gamma_{0,1}(H))Z(S_{p-n})$ is the cycle index of the group induced by $\Gamma(H)S_{p-n}$ which acts on the points of K_p and the lines of H. Each permutation

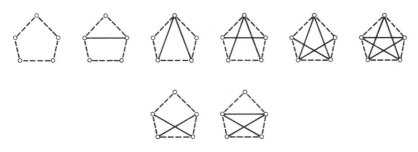

Figure 4.5.1

The eight graphs of order 5 rooted at a spanning cycle.

α in $\Gamma(H)S_{p-n}$ contributes a monomial $(\prod s_k^{j_k})(\prod t_k^{i_k})$ to $Z(\Gamma_{0,1}(H))Z(S_{p-n})$. The corresponding contribution of α to $Z(S_p)$ is denoted as usual by $(\prod s_k^{j_k})^{(2)}$. Thus, and note carefully this trick of Robinson, α contributes the product $(\prod s_k^{j_k})^{(2)}\prod s_k^{-i_k}$ to the cycle index of $\Gamma_1(\bar{H}, K_p)$ because multiplying by $\prod s_k^{-i_k}$ has the effect of eliminating the terms of $(\prod s_k^{j_k})^{(2)}$ which arise from pairs of points that are adjacent in H. Therefore we define the linear operator ρ (standing for Robinson) for point-line cycle indexes by specifying it for monomials as follows:

$$\rho((\prod s_k^{j_k})(\prod t_k^{i_k})) = (\prod s_k^{j_k})^{(2)}\prod s_k^{-i_k}. \tag{4.5.7}$$

Now the counting of supergraphs takes the form:

Theorem   The counting polynomial $r_p(x)$ that counts supergraphs of order p of any graph H of order $n \le p$ is

$$r_p(x) = Z(\Gamma_1(\bar{H}, K_p), 1 + x), \tag{4.5.8}$$

where

$$Z(\Gamma_1(\bar{H}, K_p)) = \rho(Z(\Gamma_{0,1}(H))Z(S_{p-n})). \tag{4.5.9}$$

Note that if H is a graph of order p as in (4.5.1), then (4.5.9) becomes

$$Z(\Gamma_1(\bar{H})) = \rho Z(\Gamma_{0,1}(H)). \tag{4.5.10}$$

Thus the operator ρ enables us to obtain the line cycle index for any graph from the point-line cycle index of its complement.

We illustrate with $H = C_6$, the cycle of order 6 and take $p = 6$. The point-line cycle index of any cycle graph C_p is readily computed from (2.2.11), the formula for $Z(s_p)$. When $p = 6$ we have

$$Z(\Gamma_{0,1}(C_6)) = \tfrac{1}{12}(s_1^6 t_1^6 + s_2^3 t_2^3 + 2s_3^2 t_3^2 + 2s_6 t_6 + 3s_1^2 s_2^2 t_2^3 + 3s_2^3 t_1^2 t_2^2). \tag{4.5.11}$$

On applying ρ to (4.5.11) we find

$$
\begin{aligned}
Z(\Gamma_1(\bar{C}_6)) &= \tfrac{1}{12}(s_1^{15}s_1^{-6} + s_1^3 s_2^6 s_2^{-3} + 2s_3^5 s_3^{-2} + 2s_3 s_6^2 s_6^{-1} \\
&\quad + 3s_1^3 s_2^6 s_2^{-3} + 3s_1^3 s_2^6 s_1^{-2} s_2^{-2}) \\
&= \tfrac{1}{12}(s_1^9 + 4s_1^3 s_2^3 + 2s_3^3 + 2s_3 s_6 + 3s_1 s_2^4). \tag{4.5.12}
\end{aligned}
$$

From (4.5.8) and (4.5.12), it follows routinely that

$$
\begin{aligned}
r_6(x) &= 1 + 2x + 6x^2 + 12x^3 + 16x^4 + 16x^5 + 12x^6 + 6x^7 \\
&\quad + 2x^8 + x^9. \tag{4.5.13}
\end{aligned}
$$

Of course this polynomial could also be obtained as for (4.5.2–4). The coefficient of x^2 is verified in Figure 4.5.2.

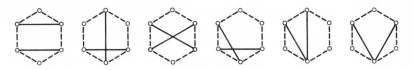

Figure 4.5.2

All six (6,8) graphs rooted at a spanning cycle.

We now turn to m-colored graphs and follow Robinson's treatment [R17] by applying Theorem (4.5.8). By definition, the points of an m-colored graph have been partitioned into m sets so that adjacent points are always in different sets. The points of each set are given the same color. For example, the 3-colored graphs in Figure 4.5.3 have two points of each color and the points joined by dashed lines have the same color.

As in Section 4.3 we determine the number of m-colored graphs for a specified m-part partition (j) of p:

$$m = \sum_{k=1}^{p} j_k, \tag{4.5.14}$$

with $\sum_{k=1}^{p} kj_k = p$ as usual. Let $b_p((j), x)$ be the polynomial that enumerates m-colored graphs of order p with (j) as color partition. For example, with $p = 6$ and partition $2 + 2 + 2$, i.e., $j_2 = 3$, $b_6((j), x)$ counts 3-colored graphs of order 6 which have two points of each color. We will show that for this partition of 6,

$$b_6((j), x) = 1 + x + 4x^2 + 9x^3 + 18x^4 + 24x^5 + 30x^6$$
$$+ 24x^7 + 18x^8 + 9x^9 + 4x^{10} + x^{11} + x^{12}. \tag{4.5.15}$$

Figure 4.5.3 verifies the coefficient of x^3 in this equation.

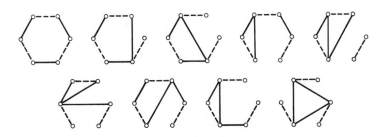

Figure 4.5.3

All nine (6,3) 3-colored graphs with two points of each color.

We define the *complete m-partite graph* $K(j)$ for any m-part partition (j) of p as the complement of the union of m complete graphs:

$$K(j) = \overline{\bigcup_{k=1}^{p} j_k K_k}. \tag{4.5.16}$$

As before, the m-colored graphs we wish to count are the spanning subgraphs of $K(j)$. Therefore it follows from (4.3.2) that

$$b_p((j), x) = Z(\Gamma_1(K(j)), 1 + x), \tag{4.5.17}$$

and we must find a method for determining the cycle index of $\Gamma_1(K(j))$. But the spanning subgraphs of the complete m-partite graph $K(j)$ are precisely the supergraphs of $\overline{K(j)}$. This suggests the idea of expressing the cycle index of $\Gamma_1(K(j))$ in terms of the operator ρ and $Z(\Gamma_{0,1}(\overline{K(j)}))$ by taking $\overline{H} = K(j)$ in (4.5.10). Thus it is sufficient to find a means for calculating the cycle index of $\Gamma_{0,1}(\overline{K(j)})$, and this we now do.

We denote by nG the graph which consists of n disjoint copies of the connected graph G. It follows [H1, p. 166] from the definition of the wreath product that

$$\Gamma(nG) = S_n[\Gamma(G)]. \tag{4.5.18}$$

Therefore from (4.3.14) the cycle index of this group is

$$Z(\Gamma(nG)) = Z(S_n)[Z(\Gamma(G))]. \tag{4.5.19}$$

Equation (4.5.19) is readily extended to obtain

$$Z(\Gamma_{0,1}(nG)) = Z(S_n)[Z(\Gamma_{0,1}(G))]. \tag{4.5.20}$$

Using the same example as in (4.5.15) by taking $G = K_2$ and $n = 3$, formula (4.5.20) becomes

$$Z(\Gamma_{0,1}(3K_2)) = Z(S_3)[Z(\Gamma_{0,1}(K_2))]. \tag{4.5.21}$$

Now $Z(\Gamma_{0,1}(K_2)) = \frac{1}{2}(s_1^2 t_1 + s_2 t_1)$, and if we substitute this expression with all subscripts multiplied by k for each s_k in $Z(S_3)$, we obtain

$$Z(\Gamma_{0,1}(3K_2)) = \tfrac{1}{48}s_1^6 t_1^3 + \tfrac{1}{16}s_1^4 s_2 t_1^3 + \tfrac{1}{16}s_1^2 s_2^2 t_1^3 + \tfrac{1}{48}s_2^3 t_1^3$$
$$+ \tfrac{1}{8}s_1^2 s_2^2 t_1 t_2 + \tfrac{1}{8}s_1^2 s_4 t_1 t_2 + \tfrac{1}{8}s_2^3 t_1 t_2$$
$$+ \tfrac{1}{8}s_2 s_4 t_1 t_2 + \tfrac{1}{6}s_3^2 t_3 + \tfrac{1}{6}s_6 t_3. \tag{4.5.22}$$

In the general case of $\overline{K(j)}$, whose components are complete graphs, the group can be expressed [H1, p. 166] as the product

$$\Gamma(\overline{K(j)}) = \prod_k S_{j_k}[\Gamma(K_k)]. \tag{4.5.23}$$

From (4.5.20) it follows that

$$Z(\Gamma_{0,1}(\overline{K(j)})) = \prod_k Z(S_{j_k})[\Gamma_{0,1}(K_k)]. \tag{4.5.24}$$

For convenience we denote $\Gamma_{0,1}(K_n)$ by $S_n^{(1,2)}$ and observe that the point-line cycle index is easily obtained from $Z(S_n^{(2)})$. As noted above, $\overline{K(j)} = 3K_2$ when $p = 6$ and $j_2 = 3$. Thus (4.5.22) provides the cycle index of the point-line group of the complement of the complete 3-partite graph of order 6 with two points of each color.

Now we can summarize Robinson's method [R17] for counting m-colored graphs.

Theorem The counting polynomial $b_p((j), x)$ for m-colored graphs of order p with partition (j) is

$$b_p((j), x) = Z(\Gamma_1(K(j)), 1 + x) \tag{4.5.25}$$

where

$$Z(\Gamma_1(K(j))) = \rho(\prod_k Z(S_{j_k})[Z(S_k^{(1,2)})]). \tag{4.5.26}$$

In the example, we have already found $Z(S_3)[Z(S_2^{(1,2)})]$ in (4.5.22). From the definition of ρ we have for $j_2 = 3$,

$$Z(\Gamma_1(K(j))) = \tfrac{1}{48}s_1^{15}s_1^{-3} + \tfrac{1}{16}s_1^7s_2^4s_1^{-3} + \tfrac{1}{16}s_1^3s_2^6s_1^{-3} + \tfrac{1}{48}s_1^3s_2^6s_1^{-3}$$
$$+ \tfrac{1}{8}s_1^3s_2^6s_1^{-1}s_2^{-1} + \tfrac{1}{8}s_1s_2s_4^3s_1^{-1}s_2^{-1} + \tfrac{1}{8}s_1^3s_2^6s_1^{-1}s_2^{-1}$$
$$+ \tfrac{1}{8}s_1s_2s_4^3s_1^{-1}s_2^{-1} + \tfrac{1}{6}s_3^5s_3^{-1} + \tfrac{1}{6}s_3s_6^2s_3^{-1}. \tag{4.5.27}$$

On simplifying this expression we have

$$Z(\Gamma_1(K(j))) = \tfrac{1}{48}s_1^{12} + \tfrac{1}{16}s_1^4s_2^4 + \tfrac{1}{12}s_2^6 + \tfrac{1}{4}s_1^2s_2^5 + \tfrac{1}{4}s_4^3 + \tfrac{1}{6}s_3^4 + \tfrac{1}{6}s_6^2. \tag{4.5.28}$$

Finally we make the substitution $s_k = 1 + x^k$ in the right side of (4.5.28) to obtain (4.5.15). Using this theorem Robinson has computed the number of 3-colored graphs for $p \le 9$ and all possible partitions.

Although the enumeration of bicolorable graphs is feasible as mentioned in Section 4.3, we must emphasize that the counting of m-colorable graphs with $m \ge 3$ remains unsolved, see Chapter 10.

4.6 BOOLEAN FUNCTIONS

A boolean function of n variables may be regarded as a mapping from the set of all n-sequences of zeros and ones into $\{0, 1\}$. In disguise, it is therefore a subset of points of the n-cube Q_n. Our object is to count the number of types of boolean functions with equivalence determined by the group of Q_n.

We have seen that the cycle index of the line-group of G serves to count spanning subgraphs. In this section we study the uses of the cycle index of the point-group of G and find that it is useful in counting dissimilar sets of points. The most interesting application of this technique enables us to enumerate the types of boolean functions.

By way of preliminaries for the counting of boolean functions, we interpret equation (2.5.1) for $Z(A, 1 + x)$ in the case that A is the group $\Gamma(G)$ of a given graph G.

(4.6.1)   **Theorem**   The coefficient of x^r in $Z(\Gamma(G), 1 + x)$ is the number of dissimilar sets of r points, with equivalence determined by the automorphism group of G.

To illustrate, consider the graph G of Figure 4.3.1. The cycle index of its group is given by

$$Z(\Gamma(G)) = [\tfrac{1}{2}(s_1^2 + s_2)]^2 = \tfrac{1}{4}(s_1^4 + 2s_1^2 s_2 + s_2^2) \qquad (4.6.2)$$

so that

$$Z(\Gamma(G), 1 + x) = 1 + 2x + 3x^2 + 2x^3 + x^4. \qquad (4.6.3)$$

The middle term of this polynomial is verified by Figure 4.6.1 in which the two points in the subset are solid.

Pólya [P9] found that the early work of Jevons [J1] and Clifford [C4] on counting types of boolean functions led to numerical results which were not entirely correct. The problem was to find the number of "different" propositions composed of n statements, each of which has two truth values 0 or 1. Using the familiar symbols $\vee$ (or) and $\wedge$ (and) for disjunction and conjunction, the proposition $P_1 \vee (P_2 \wedge P_3)$ is composed of P_1, P_2, and P_3. Now with 1 and 0 representing the two truth values, one sees that composed propositions correspond in a natural way to boolean functions. For example, $P_1 \vee (P_2 \wedge P_3)$ corresponds to the function f which sends $(0, 0, 0)$, $(0, 0, 1)$, and $(0, 1, 0)$ to 0 and all other triples to 1. But we wish to consider two composed propositions to be equivalent not only when their corresponding

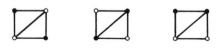

Figure 4.6.1

The dissimilar sets of two points in a random graph.

boolean functions are identical, but also whenever one can be obtained from the other by permuting the statement or changing any statements to their negation. Thus it is seen that $P_1 \vee (P_2 \wedge P_3)$ is considered equivalent to $(P_3 \vee \text{not } P_1) \wedge (P_3 \vee P_2)$ by using the distributive law for disjunction over conjunction.

Pólya recognized that the problem at hand was to determine the number of ways in which the 2^n points of the n-cube Q_n can be colored using two different colors. This follows from the fact that the group of Q_n is identical to the group that permutes the n statements arbitrarily while also changing statements to their negation. We can express this group using the exponentiation operation [H9] so that

$$\Gamma(Q_n) = [S_2]^{S_n}. \tag{4.6.4}$$

The exponent group S_n indicates that the n propositions P_i can be permuted at will, while the base group S_2 allows for the possibility of negating each variable after this permutation. On applying (4.6.1) with this permutation group, the main result of this section can be stated. The cycle index of this group was first presented by Slepian [S3] in a rather involved appearance which did not recognize the exponentiation group as such.

(4.6.5)   **Theorem**   The number $N(n, r)$ of different boolean functions of n variables which have exactly r nonzero values is the coefficient of x^r in $Z([S_2]^{S_n}, 1 + x)$.

To illustrate, we display in Figure 4.6.2 the six ways of coloring the points of Q_3 with four points of each color, thus verifying the coefficient of x^6 in $Z([S_2]^{S_3}, 1 + x)$. Note that the necklaces of Figure 2.4.2 correspond precisely to boolean functions of two variables, because in a four-bead necklace, the action takes place on the 2-cube.

Using the fact that $\Gamma(Q_n)$ and the wreath product $S_n[S_2]$ are abstractly isomorphic. Pólya calculated the cycle indexes $Z(\Gamma(Q_n))$ for $n \leq 4$. On applying the theorem above, he found the entries in Table 4.6.1. Because of the close connection between boolean algebras and switching circuits, Slepian [S3] was motivated to calculate the total number of boolean function

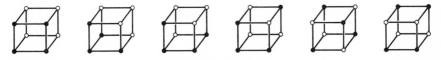

Figure 4.6.2

The cubes with four points of each color.

TABLE 4.6.1

THE NUMBER OF BOOLEAN FUNCTIONS

n \ r	0	1	2	3	4	5	6	7	8	Total
1	1	1	1							3
2	1	1	2	1	1					6
3	1	1	3	3	6	3	1	1	1	22
4	1	1	4	6	19	27	50	56	74	402

for $n = 5$ and 6. Harrison and High [HH1] also devised a rather complicated method for determining $Z([B]^{S_n})$ and used it for computing some classes of Post functions. The most compact formulas for $Z([B]^4)$ were found by Robinson and Palmer as reported in [P3]. We now sketch some of the details of the latter approach in order to supply a reasonable means for obtaining the cycle index in Theorem (4.6.5).

First we define a product, denoted by $\times$ for monomials, compare (4.1.16), by

$$\prod s_k^{j_k} \times \prod s_k^{i_k} = \prod_{r,t} s_{[r,t]}^{(r,t)j_r i_t}. \qquad (4.6.6)$$

As shown in [H7], if this product is extended linearly, the cycle index $Z(A \times B)$ of the cartesian product can be expressed as

$$Z(A \times B) = Z(A) \times Z(B). \qquad (4.6.7)$$

For each positive integer r, a new operator J_r is defined for monomials by

$$J_r\left(\prod_{k=1}^{n} s_k^{j_k}\right) = \prod_{v=1}^{n^r} s_v^{i_v} \qquad (4.6.8)$$

where

$$\sum_{w|v} w i_w = \left(\sum_{k|v/(r,v)} k j_k\right)^{(r,v)}. \qquad (4.6.9)$$

Clearly J_1 is the identity operator. Using möbius inversion on (4.6.9), we can write

$$i_v = \frac{1}{v} \sum_{w|v} \mu(v/w)\left(\sum_{k|w/(r,w)} k j_k\right)^{(r,w)}. \qquad (4.6.10)$$

The J_r are then extended linearly. The preceding five formulas can be used to express $Z([B]^4)$ in terms of $Z(A)$ and $Z(B)$ for any permutation groups A and B. The manner in which this is accomplished will be clear when we illustrate with $A = S_3$ and $B = S_2$, so that we are dealing with the group of Q_3. The hardest part is to calculate the operators J_r. First we substitute the operator J_r for each variable s_r in $Z(S_3)$ to obtain

$$Z(S_3; J_1, J_2, J_3) = \tfrac{1}{6}(J_1^3 + 3J_1J_2 + 2J_3). \tag{4.6.11}$$

Next, the terms of (4.6.11) act on $Z(S_2)$ as follows:

$$J_1^3(Z(S_2)) = J_1(Z(S_2)) \times J_1(Z(S_2)) \times J_1(Z(S_2))$$

$$J_1J_2(Z(S_2)) = J_1(Z(S_2)) \times J_2(Z(S_2)) \tag{4.6.12}$$

It follows from the definitions that

$$J_1(Z(S_2)) = Z(S_2) = \tfrac{1}{2}(s_1^2 + s_2) \tag{4.6.13}$$

and

$$J_2(Z(S_2)) = \tfrac{1}{2}(J_2(s_1^2) + J_2(s_2)) = \tfrac{1}{2}(s_1^2 s_2 + s_4) \tag{4.6.14}$$

and

$$J_3(Z(S_2)) = \tfrac{1}{2}(J_3(s_1^2) + J_3(s_2)) = \tfrac{1}{2}(s_1^2 s_3^2 + s_2 s_6). \tag{4.6.15}$$

From (4.6.12) and the definition (4.6.7) of the cartesian product $\times$ for polynomials, we find

$$J_1^3(Z(S_2)) = \frac{1}{2^3}(s_1^8 + 7s_2^4), \qquad J_1J_2(Z(S_2)) = \frac{1}{2^2}(s_1^4 s_2^2 + s_2^4 + 2s_4^2). \tag{4.6.16}$$

Lastly, on substituting the polynomials of (4.6.15) and (4.6.16) for the corresponding terms in (4.6.11), we obtain $Z([S_2]^{S_3})$ in a form reminiscent of the equations in Pólya [P9]:

$$Z([S_2]^{S_3}) = \frac{1}{6}\left\{ \frac{(s_1^8 + 7s_2^4)}{2^3} + 3\frac{(s_1^4 s_2^2 + s_2^4 + 2s_4^2)}{2^2} + 2\frac{(s_1^2 s_3^2 + s_2 s_6)}{2} \right\} \tag{4.6.17}$$

Thus as shown in [P3], the cycle index $Z([B]^A)$ of any exponentiation group can be expressed as the image of $Z(B)$ under the operator $Z(A; J_1, J_2, \ldots)$. In this way, all enumeration problems in which the configuration group can be viewed as an exponentiation group can be solved in principle.

4.7 EULERIAN GRAPHS

Even graphs (with every point of even degree) and eulerian graphs (connected even graphs) were counted in Section 1.4 *in the labeled case*. The more subtle problem for unlabeled graphs was solved when Robinson [R18] viewed this challenge from just the proper perspective. To count eulerian graphs with p points, he first counted the even ones and then applied the usual technique for expressing the number of connected configurations in terms of the total number.

This section consists mainly of the proof of Robinson's formula for the number of even graphs of order p. The formula bears a noticeable resemblance to that for counting graphs, using the cycle index (4.1.9) of the pair group of the symmetric group, for it is based on that result.

Theorem The number w_p of even graphs of order p is

$$w_p = \frac{1}{p!} \sum_{(j)} \frac{p!}{\prod_k k^{j_k} j_k!} 2^{e(j)} \tag{4.7.1}$$

where sgn is the usual sign function in

$$e(j) = \sum_{r < t} (r, t) j_r j_t + \sum_{k=1} k \binom{j_k}{2}$$

$$+ \sum_{k=0} (k - 1)(j_{2k} + j_{2k+1}) + \text{sgn} \left(\sum_{k=0} j_{2k+1} \right). \tag{4.7.2}$$

Proof Following Robinson, we apply the variant (2.3.9) of Burnside's Lemma in which there is a restriction to the appropriate subset of the objects. We have seen that the orbits of the power group $E_{2^p}^{S_p^{(2)}}$ correspond to graphs of order p. On restricting this group to the set F of functions that represent even graphs, we can apply (2.3.9) to express w_p in terms of the functions in F fixed by the permutations in this power group. In other words for each permutation α in S_p, we seek the number of functions f in F such that

$$f\{i, j\} = f\{\alpha i, \alpha j\}. \tag{4.7.3}$$

Thus to prove the theorem, we must show that if the cycles of α determine the partition (j) of p, then the number of functions in F satisfying (4.7.3) is $2^{e(j)}$ where $e(j)$ is given by (4.7.2).

Before launching into the details of this proof, we try to clarify the treatment by displaying in Figure 4.7.1 an even graph which corresponds to a function f in F satisfying (4.7.3) for a specific permutation α of degree 13, namely

$$\alpha = (1\ 2\ 3\ 4)(5\ 6)(7\ 8\ 9)(10)(11)(12)(13).$$

By only a slight abuse of language, we say that this graph is fixed by α. Both solid and dashed lines appear in order to see the action of α on the lines

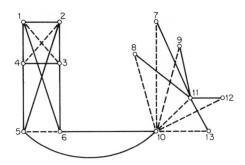

Figure 4.7.1

An even graph fixed by $\alpha = (1\ 2\ 3\ 4)(5\ 6)(7\ 8\ 9)(10)(11)(12)(13).$

more vividly. The sense in which some lines are solid and others are dashed will hopefully become clear as the proof unfolds.

First we find the contribution to $2^{e(j)}$ made by α when the cycles of α are considered individually. If $z_k = (123 \cdots k)$ is any odd cycle of α, then we know that z_k induces $(k - 1)/2$ cycles of 2-subsets of its points. Each point of z_k is incident with exactly two lines of each of these cycles, for example, see Figure 4.7.2a. Hence each of these $(k - 1)/2$ cycles may be either included or excluded in an even graph fixed by α. Therefore the contribution to $e(j)$ for all odd cycles considered individually is $\sum_{k=1}^{P} j_k(k - 1)/2$; compare (4.1.14). Similarly if k is even, the contribution to $e(j)$ is $\sum_{k=2}^{P} j_k(k - 1)/2$; compare (4.1.15). Note that the cycle of 2-subsets of length $k/2$ is excluded because each point of z_k is incident with only one of the 2-subsets of this cycle. For example, in Figure 4.7.2b, this cycle is shown with dashed lines and it contributes an odd degree (one) to the points in the figure.

As in counting graphs we next consider pairs of cycles z_r and z_t of α and the 2-subsets which have one point in each of these cycles. In determining

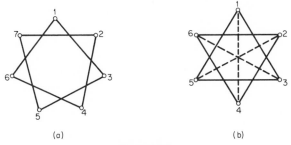

(a)   (b)

Figure 4.7.2

Graphs fixed by odd and even cycles.

the cycle index of the pair group, we saw that these cycles z_r and z_t induce (r, t) cycles of length $[r, t]$ on such 2-subsets; compare (4.1.16). Thus each point of z_r is in $[r, t]/r$ of the 2-subsets of one such cycle of length $[r, t]$. Note that if t is even and r is odd, each point of z_r is in an even number of such 2-subsets. With this in mind, we now consider all pairs of cycles z_r and z_t with r or t even. Let U be any collection of cycles of 2-subsets with one point in each of these two different cycles z_r, z_t. Then each point of an odd cycle is in an even number of 2-subsets in the cycles of U. Each point of an even cycle z_k may be in either an even or an odd number of these 2-subsets. Therefore we see that the cycles of U can determine an even graph by either including or excluding the cycle of length $k/2$ induced by z_k, as necessary.

Adding the two cases $r \neq t$ and $r = t$, we see that the total number of cycles in U is

$$\sum_{r < t} (r, t) j_r j_t + \sum_{k=1} 2k \binom{j_{2k}}{2},$$

where at least one of r and t is even. Hence this is the contribution to $e(j)$ from pairs of objects between two cycles of α, at least one of which is even. Note that in Figure 4.7.1 it is necessary to include the dashed lines of the cycle (1234) as well as the dashed line of the cycle (56), in order to get an even graph.

We have not yet handled the 2-subsets with one point in each of two different odd cycles. We select one special odd cycle, say z, and let W be a collection of cycles of 2-subsets obtained by choosing for each odd cycle $z_r \neq z$ one of the cycles of 2-subsets whose points are in z and z_r. Thus W will have $(\sum_{k=0} j_{2k+1}) - 1$ cycles. We can now determine an even graph for each selection of those cycles of 2-subsets not in W. This is accomplished by either including or excluding each cycle in W as necessary to make each point of the odd cycles other than z lie in an even number of 2-subsets. Then, since Euler proved that the number of points of odd degree in any graph must be even, the points of z as well will be in an even number of 2-subsets. The number of such cycles not in W when r and t are both odd is

$$\sum_{r < t} (r, t) j_r j_t + \sum_{k=0} (2k + 1) \binom{j_{2k+1}}{2} - (m - 1)$$

where m is the number of odd cycles unless that number is 0 in which case $m = 1$. In Figure 4.7.1 we have illustrated this with $z = (10)$, and here W consists of the cycles of dashed lines emanating from the point 10. After joining point 11 with points 7, 8, 9, 12, and 13, it was necessary in this case to include all the lines of W to obtain an even graph.

Combining all these observations and adding the various contributions, (4.7.2) is verified, and even graphs are counted.   //

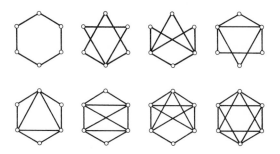

Figure 4.7.3

The eight eulerian graphs of order 6.

Robinson used these formulas to calculate the number w_p of even graphs for $p \leq 8$:

$$w(x) = x + x^2 + 2x^3 + 3x^4 + 7x^5 + 16x^6 + 54x^7 + 283x^8 + \cdots. \quad (4.7.4)$$

The seven even graphs of order 5 are shown in Figure 1.4.3, where labeled even graphs were counted, verifying the coefficient of x^5. The generating function $u(x)$ for eulerian graphs can be obtained in the usual way by again applying (4.2.3), which counts connected graphs in terms of all graphs. The first few terms of $u(x)$ are

$$u(x) = x + x^3 + x^4 + 4x^5 + 8x^6 + 37x^7 + 184x^8 + \cdots. \quad (4.7.5)$$

The coefficient of x^6 is verified in Figure 4.7.3.

EXERCISES

4.1 Find the polynomial which counts the ways in which the lines of (a) the (5, 9) graph and (b) $K_{3,3}$ can be colored with three colors. (*Hint*: See (4.3.2).)

4.2 Find the cycle indexes $Z(S_2[C_3])$, $Z(C_3[S_2])$, and $Z(S_3[D_4])$.

4.3 Find the cycle index for the graph with four components, (a) each a triangle, (b) two triangles and two isolated points.

4.4 *Pure two-dimensional simplicial complexes*, which consist of a finite set X together with a collection of 3-subsets of X.   (Harary [H4])

4.5 Multigraphs of *strength s*, in which at most s lines join any pair of points.

4.6 *Signed graphs*, in which the lines may be positive or negative.

(Harary [H3])

4.7 *Balanced* signed graphs, in which no cycle has an odd number of negative lines.   Harary and Palmer [HP10])

4.8 *Point-symmetric* graphs, whose groups are transitive, of order p, a prime.

(Turner [T1])

Lord Ronald jumped on his horse and rode off madly in all directions.

Stephen Leacock

Chapter 5 DIGRAPHS

In this chapter, the four sections count digraphs, tournaments, orientations of a given graph, and "mixed graphs" in which both arcs and undirected lines can appear. In addition to enumerating all digraphs, we handle relations and connected digraphs. The counting of tournaments yields that of strong tournaments as a corollary. Our formula for the number of mixed graphs contains as special cases, which easily can be stated explicitly, the counting of graphs, digraphs, oriented graphs, complete digraphs, and tournaments.

A word about enumeration methods is in order. Digraphs are counted by modifying only slightly Pólya's method for enumerating graphs. This is accomplished by developing a new unary operation on a permutation group, called the reduced ordered pair group, whose purpose is to act on ordered pairs of distinct objects involved in the original permutation group.

The cycle index of the reduced ordered pair group of S_p is used to count digraphs. This cycle index is then modified to enable us to count tournaments, by restricting the associated power group. Applying this idea to the reduced ordered pair group of an arbitrary graph and using two sets of variables in the resulting cycle index, we can then count its orientations. These methods culminate in a general theorem for counting mixed graphs, which makes use of the same cycle index as for digraphs, again with two sets of variables and special purpose figure counting series.

5.1 DIGRAPHS

This section parallels the derivation of the graph-counting polynomial $g_p(x)$ and the associated cycle index of the pair group of S_p. While there are only 11 graphs of order 4, there are 218 digraphs of order 4. The formidable task of drawing all 218 digraphs has been performed, and the diagrams may be found in Appendix 2 of [H1, p. 226].

Let $d_p(x)$ be the counting polynomial for digraphs of order p so that, for example

$$d_3(x) = 1 + x + 4x^2 + 4x^3 + 4x^4 + x^5 + x^6. \tag{5.1.1}$$

These coefficients are verified by Figure 1.1.5.

We now show, following [H11], how PET can be used to determine $d_p(x)$.

Let $X = \{1, 2, \ldots, p\}$, and let $X^{[2]}$ denote all ordered pairs of *different elements* of X. With $Y = \{0, 1\}$, the functions from $X^{[2]}$ into Y represent labeled digraphs of order p. Each function f corresponds to that digraph $D(f)$ with point set X in which there is an arc from the point i to the point j if and only if $f(i, j) = 1$. Thus two functions f and g represent the same digraph if there is a permutation α of X such that if i is adjacent to j in $D(f)$ then αi is adjacent to αj in $D(g)$. Therefore $D(f)$ and $D(g)$ are isomorphic if and only if

$$f(i, j) = g(\alpha i, \alpha j) \tag{5.1.2}$$

for all (i, j) in $X^{[2]}$.

This equation suggests the next unary operation on permutation groups. If A is a permutation group with object set X, the *reduced ordered pair group* of A, denoted $A^{[2]}$, has $X^{[2]}$ as its object set and is induced by A. That is, for each α in A, there is a permutation α' in $A^{[2]}$ such that for every pair (i, j) in $X^{[2]}$ the image under α' is given by

$$\alpha'(i, j) = (\alpha i, \alpha j). \tag{5.1.3}$$

We can now state the formula for the number of digraphs.

Theorem The counting polynomial $d_p(x)$ for digraphs of order p is given by

$$d_p(x) = Z(S_p^{[2]}, 1 + x) \tag{5.1.4}$$

where

$$Z(S_p^{[2]}) = \frac{1}{p!} \sum_{(j)} \frac{p!}{\prod k^{j_k} j_k!} \prod_{k=1}^{p} s_k^{(k-1)j_k + 2k(\frac{j_k}{2})} \prod_{r<t} s_{[r,t]}^{2(r,t)j_r j_t}. \qquad (5.1.5)$$

Proof The equation (5.1.4) is in the form of Corollary (2.5.1) of the PET. We show that the reduced ordered pair group of S_p is the correct configuration group. Two functions f and g from $X^{[2]}$ into $\{0, 1\}$ represent the same digraph if and only if there is a permutation α in S_p such that

$$f(z) = g(\alpha'z) \qquad (5.1.6)$$

for every z in $X^{[2]}$. Consequently, the orbits of the power group $E_2^{S_p^{[2]}}$ correspond to the different digraphs.

The figure counting series is $1 + x$ because, as for graphs, the arc (i, j) is either absent or present. Thus by the usual definitions (2.4.4) of the weights of functions and orbits, the weight of an orbit of $E_2^{S_p^{[2]}}$ is the number of arcs in the corresponding digraph, proving (5.1.4).

Our object now is to derive formula (5.1.5) for $Z(S_p^{[2]})$, but first we illustrate with $p = 4$ in order to clarify the derivation of the more general cycle index. Each of the five permutations in S_4 induces a permutation in the reduced ordered pair group $S_4^{[2]}$. Table 5.1.1 shows the terms in $S_4^{[2]}$ induced by each term of $Z(S_4)$.

One of these terms is now illustrated. In Figure 5.1.1 the ordered pairs are indicated for conciseness by juxtaposition: $(1, 2) = 12$, and so on.

Collecting terms, we find that

$$Z(S_4^{[2]}) = \tfrac{1}{24}(s_1^{12} + 6s_1^2 s_2^5 + 8s_3^4 + 3s_2^6 + 6s_4^3), \qquad (5.1.7)$$

so

$$Z(S_4^{[2]}, 1 + x) = d_4(x) = 1 + x + 5x^2 + 13x^3 + 27x^4 + 38x^5 + 48x^6$$
$$+ 38x^7 + 27x^8 + 13x^9 + 5x^{10} + x^{11} + x^{12}.$$

$$(5.1.8)$$

TABLE 5.1.1

Term of $Z(S_4)$	Induced term of $Z(S_4^{[2]})$
s_1^4	s_1^{12}
$s_1^2 s_2$	$s_1^2 s_2^5$
$s_1 s_3$	s_3^4
s_2^2	s_2^6
s_4	s_4^3

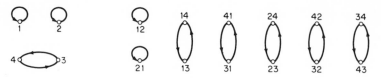

Figure 5.1.1

The permutation $(1)(2)(3\ 4)$ and its induced permutation in $S_4^{[2]}$.

We indicate the correspondence between terms of the cycle indexes of S_p and $S_p^{[2]}$ for $p = 4$ by writing

$$s_1^4 \rightarrow s_1^{12}, \qquad s_1^2 s_2 \rightarrow s_1^2 s_2^5$$

and so on. To obtain an expression for $Z(S_p^{[2]})$, we need to find the missing right hand member of

$$s_1^{j_1} s_2^{j_2} \cdots s_p^{j_p} \rightarrow ? \tag{5.1.9}$$

Just as in the derivation of $Z(S_p^{(2)})$, when we consider more than one cycle in the same permutation α of S_p, there are two separate contributions to the corresponding term of $Z(S_p^{[2]})$. The first comes from pairs of elements of X, both in a common cycle of α; the second from pairs of elements, one in each of two different cycles of α.

To determine the first of these contributions, consider a cycle z_k of length k in α. There are $k(k - 1)$ pairs in $X^{[2]}$ of elements permuted by z_k. These pairs are permuted in cycles of length k by α'. Hence α' permutes them in $k - 1$ cycles of length k each. Thus if there are j_k cycles of length k, the pairs of elements lying in common cycles contribute

$$s_k^{j_k} \rightarrow s_k^{(k-1)j_k}. \tag{5.1.10}$$

Now consider any two cycles of α, say z_r and z_t of length r and t respectively. There are $2rt$ pairs (i, j) in $X^{[2]}$, with i permuted by z_r and j permuted by z_t. These pairs are permuted in cycles of length $[r, t]$ by α'. Hence α' permutes them in $2(r, t)$ cycles of length $[r, t]$ each. Thus the contribution of pairs of cycles of length r with cycles of different length $t \neq r$ is

$$s_r^{j_r} s_t^{j_t} \rightarrow s_{[r,t]}^{2(r,t)j_r j_t}. \tag{5.1.11}$$

Finally the contribution of pairs of cycles of the same length k is

$$s_k^{j_k} \rightarrow s_k^{2k\binom{j_k}{2}}. \tag{5.1.12}$$

Now the right side of (5.1.9) and hence also of equation (5.1.5) can be filled in by multiplying the right sides of (5.1.10), (5.1.11), and (5.1.12) over all applicable indexes.   //

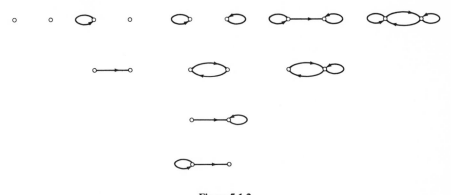

Figure 5.1.2

The relations on two points.

The formulas in this theorem were used to compute the number d_p of digraphs of order $p \leq 8$ in Appendix I and in Table 5.1.1. The results agree with those of Oberschelp [O1], who used the same formulas.

No loops are permitted in digraphs since they are defined as irreflexive relations. To count relations on p points in which loops can occur, we need to extend the object set of $S_p^{[2]}$ to include *all* ordered pairs in $X \times X$. The permutation group obtained is called the *ordered pair group* of S_p and is denoted by S_p^2. Its cycle index is obtained on modifying formula (5.1.5) for $Z(S_p^{[2]})$ by multiplying each term in the sum by $\prod s_k^{jk}$. Then the polynomial $r_p(x)$ which counts relations is given by

$$r_p(x) = Z(S_p^2, 1 + x). \tag{5.1.13}$$

From this equation, we find that for $p = 2$

$$r_2(x) = 1 + 2x + 4x^2 + 2x^3 + x^4, \tag{5.1.14}$$

and these coefficients are verified in Figure 5.1.2.

A digraph is called *connected* if on ignoring all orientations on its arcs, the resulting multigraph is connected. Such digraphs are called "weakly connected" in [HNC1]. Connected digraphs are counted in terms of all digraphs by the identical method (4.2.3) used for connected graphs. We conclude by observing that if g_p in formula (4.2.6) is replaced by the number d_p of digraphs of order p, then equation (4.2.9) which involves parameters a_p can be used to compute the number of connected digraphs. The results of this computation for $p \leq 8$ are in Table 5.1.2.

TABLE 5.1.2

CONNECTED DIGRAPHS

p	d_p	pa_p	c_p
1	1	1	1
2	3	5	2
3	16	40	13
4	218	801	199
5	9 608	46 821	9 364
6	1 540 944	9 185 102	1 530 843
7	882 033 440	6 163 297 995	880 471 142
8	1 793 359 192 848	14 339 791 693 249	1 792 473 955 306

5.2 TOURNAMENTS

Labeled tournaments (Figure 1.1.6) are equal in number to labeled graphs; this is emphatically not so for unlabeled tournaments. All tournaments with $p = 2, 3$, and 4 points are shown in Figure 5.2.1. The first tournament with three points is called a *transitive triple*; the second a *cyclic triple*. For a concise, comprehensive treatment of topics on tournaments, the book [M2] of Moon is an excellent source.

The counting of tournaments is due to Davis [D1]; see also Moon [M2, p. 84].

Theorem   The number $T(p)$ of tournaments of order p is

$$T(p) = \frac{1}{p!} \sum_{(j)} {}^* \frac{p!}{\prod k^{j_k} j_k!} 2^{t(j)} \qquad (5.2.1)$$

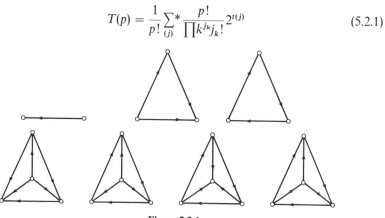

Figure 5.2.1

The smallest tournaments.

where the asterisk on $\sum$ calls attention to the unconventional summing only over those partitions (j) with $j_k = 0$ whenever k is even, and where

$$t(j) = \frac{1}{2}\left(\sum_{m,n=1}^{p} j_m j_n (m, n) - \sum_{k=1}^{p} j_k \right). \tag{5.2.2}$$

Proof We first verify (5.2.1). As in the proof of Theorem (5.1.4), the orbits of the power group $E_2^{S_p^{[2]}}$ correspond to digraphs of order p. On restricting this group to the set F of functions f which represent tournaments, namely those f for which $f(i, i) \neq f(j, i)$, the version (2.3.9) of Burnside's Lemma can be applied. As a result, $T(p)$ can be expressed in terms of the number of functions in F fixed by the permutations in the same power group. Thus for each permutation α in S_p, we seek the number of functions f in F such that

$$f(i, j) = f(\alpha i, \alpha j), \tag{5.2.3}$$

for all (i, j) in $X^{[2]}$. The functions which satisfy (5.2.3) are precisely those fixed by the permutation in this power group induced by α. By a slight abuse of language, we say that they are "fixed by α." Therefore if the cycles of α determine the partition (j) of p, we need to show that the number of functions in F fixed by α is $2^{t(j)}$.

Again let α' be the permutation in $S_p^{[2]}$ induced by α. We define the *converse* of any given cycle z' in the disjoint cycle decomposition of α' as that cycle of α' which permutes all ordered pairs (i, j) such that (j, i) is permuted by z'. A cycle z' of α' is called *self-converse* if (i, j) is permuted by z' whenever (j, i) is.

To determine the contribution to $t(j)$ made by α when the cycles of α are considered individually, let f represent a tournament fixed by α so that (5.2.3) holds. Therefore f is constant on the cycles of α'.

We now justify the asterisk on the summation sign in (5.2.1). If $z_k = (123 \cdots k)$ is any even cycle of α then α' has a self-converse cycle z', namely the one which acts on both $(1, (k/2) + 1)$ and $((k/2) + 1, 1)$. But if f is constant on this self-converse cycle z', f is not a tournament. Thus there are no tournaments fixed by α if α has any even cycles, and (5.2.1) is established.

To complete the proof of the theorem, it remains to demonstrate (5.2.2). If z_k is an odd cycle, it induces $k - 1$ cycles of pairs in $X^{[2]}$. These $k - 1$ cycles consist of $(k - 1)/2$ pairs of converse cycles. Recall that any tournament f fixed by α is constant on the cycles of α'. Therefore the elements of exactly one cycle of each of these $(k - 1)/2$ pairs must be included in a tournament fixed by α. Then the contribution to $t(j)$ due to all the odd cycles of α is $\sum j_k (k - 1)/2$ summed over odd k.

Now we consider two cycles z_m and z_n of α and the pairs in $X^{[2]}$ which have one point in each. In determining the cycle index $Z(S_p^{[2]})$, we saw that two such cycles induce $2(m, n)$ cycles of objects (ordered pairs) in $X^{[2]}$. These latter cycles consist of (m, n) pairs of converse cycles. The objects in just one of each of these pairs of converse cycles must be included in a tournament fixed by α. Therefore the contribution to $t(j)$ of all such pairs z_m and z_n with $m \neq n$ is $\sum_{m<n} j_m j_n (m, n)$. The contribution of pairs of cycles of the same length is $\sum k \binom{j_k}{2}$. On adding these three contributions we have formula (5.2.2) for $t(j)$.　　//

The values of $T(p)$ with $p \leq 12$ taken from [M2, p. 87] are displayed in Table 5.2.1.

TABLE 5.2.1

THE NUMBER OF TOURNAMENTS

p	$T(p)$
1	1
2	1
3	2
4	4
5	12
6	56
7	456
8	6 880
9	191 536
10	9 733 056
11	903 753 248
12	154 108 311 168

A digraph is *strongly connected* or *strong* if every pair of points are mutually reachable by directed paths. It is well known [HNC1, p. 306] that a tournament is strong if and only if it is hamiltonian, that is, it contains a spanning directed cycle. The strong tournaments with $p = 5$ points are now shown, using the clever device of Moon [M2, p. 92] that whenever an arc is missing it is oriented according to gravity from the higher point to the lower one.

The number of strong tournaments is easily expressed in terms of the total number of tournaments by means of generating functions; Moon [M2, p. 88]. The proof requires a few definitions. A *strong component* of a digraph D is a maximal strong subgraph. The *condensation* D^* of D has the strong components of D as its points, with its arcs induced from those of D. As noted in [HNC1, p. 298], the condensation of any tournament is a transitive tournament, that is, a complete order.

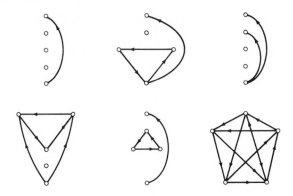

Figure 5.2.2

The strong tournaments of order 5.

Theorem Let $T(x)$ and $S(x)$ be the ordinary generating functions for tournaments and strong tournaments respectively. Then

$$S(x) = \frac{T(x)}{1 + T(x)}.$$ (5.2.4)

Proof To obtain $S(x)$, let $T_n(x)$ be the ordinary generating function for tournaments with exactly n strong components. Then by the above observations, $T_n(x) = S^n(x)$ and $T(x) = \sum T_n(x)$, proving (5.2.4). //

5.3 ORIENTATIONS OF A GRAPH

An *orientation of a graph G* is an assignment of an arrow to every line of G, i.e., an ordering of its point pair. An *oriented graph* is a digraph in which no pair of points are joined by a symmetric pair of arcs. Thus every orientation of a graph is an oriented graph. Every graph G has an invariant $o(G)$, the number of different orientations of G. One quickly verifies that $o(C_3) = 2$, $o(C_4) = o(C_5) = 4$, and $o(C_6) = 9$.

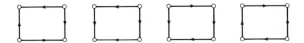

Figure 5.3.1

The orientations of the cycle of order 4.

Our purpose here is to develop an enumeration method [HP2] for determining the number $o(G)$ for a given graph G. To do this we need only generalize some of the observations made in the previous section. There we found the number $T(p)$ of tournaments of order p, which is simply $o(K_p)$, the number of orientations of the complete graph.

As in the case of K_p, a formula for $o(G)$ can be obtained by applying Burnside's Lemma. The expression obtained can be simplified by using a special cycle index in two variables. Suppose $\Gamma(G)$ acts on $X = \{1, 2, \ldots, p\}$ as usual. It is convenient to define *the digraph $D(G)$ of a graph G*. The points of $D(G)$ are those of G and the set U of arcs of $D(G)$ are all ordered pairs (i, j) such that points i and j are adjacent in G. Thus any orientation of G is obtained by choosing just one from each pair of symmetric arcs in $D(G)$.

The reduced ordered pair group $\Gamma(G)^{[2]}$ acts on *all* ordered pairs (i, j) with i and j in X and $i \neq j$, whether or not i and j are adjacent in G. We thus require the restricted permutation group $\Gamma(G)^{[2]}|U$ which acts only on the arcs in U, and denote its cycle index by $Z(\Gamma(G)^{[2]}|U)$. To distinguish between the kinds of cycles involved here, it is convenient to use the variables t_k for the self-converse cycles and s_k for the pairs of converse cycles. Thus by definition the polynomial $Z(\Gamma(G)^{[2]}|U; s_k, t_k)$ displays this distinction. Note that as in the case of tournaments, the variables t_k arise only from even cycles of length k in the automorphisms of G. The next result follows from the Restricted Form (2.3.9) of Burnside's Lemma.

Theorem   The number of orientations of a graph G is

$$o(G) = Z(\Gamma(G)^{[2]}|U; \sqrt{2}, 0). \tag{5.3.1}$$

Note that the radical $\sqrt{2}$ disappears on substitution because converse cycles occur in pairs. Furthermore, the rather unusual figure counting series consisting entirely of 0 serves the purpose of making quite certain that no symmetric pair of arcs can occur.

Because of the intractibility of the cycle index involved in (5.3.1), we can calculate the number $o(G)$ in principle only for an arbitrary graph G. However, an exact formula can be obtained when $G = C_p$, the cycle of order p. Since $\Gamma(C_p) = \Gamma_1(C_p) = D_p$, the dihedral group, we can modify formula (2.2.11) for $Z(D_p)$ to obtain $Z(\Gamma(C_p)^{[2]}|U; s_k)$. Since the variables t_k are contributed only by reflections, we have

$$Z(\Gamma(C_p)^{[2]}|U; s_k, t_k) = \frac{1}{2p} \sum_{d|p} \varphi(d) s_d^{2p/d} + \begin{cases} \frac{1}{2} s_2^{p-1} t_2 & \text{if } p \text{ is odd} \\ \frac{1}{4}(s_2^p + s_2^{p-2} t_2^2) & \text{if } p \text{ is even.} \end{cases} \tag{5.3.2}$$

Hence we can write

$$o(C_p) = \frac{1}{2p} \sum_{d|p} \varphi(d) 2^{2^{p/d}} + \begin{cases} 0 & \text{if } p \text{ is odd} \\ 2^{(p-4)/2} & \text{if } p \text{ is even.} \end{cases} \tag{5.3.3}$$

Our approach can also be modified to count oriented graphs [H6], but this result will arise as a special case of the enumeration of mixed graphs which is given in the next section.

5.4 MIXED GRAPHS

A *mixed graph* can contain both ordinary and oriented lines. For example, the graph in Figure 5.4.1 is a mixed graph with two ordinary and three oriented lines. An ordinary graph may be regarded as a mixed graph with no oriented lines, and an oriented graph as a mixed graph with no ordinary lines. Further, any digraph may be considered as a mixed graph by changing each symmetric pair of lines to an ordinary line.

Our object is to derive a formula which enumerates mixed graphs on p points with respect to both the number of ordinary and oriented lines [HP7]. Let m_{pqr} be the number of mixed graphs with p points having q oriented lines and r ordinary lines. Then the polynomial $m_p(x, y)$ which enumerates mixed graphs with p points according to both the number of ordinary and oriented lines is defined by

$$m_p(x, y) = \sum_{q,r} m_{pqr} x^q y^r, \tag{5.4.1}$$

where $q + r \leq \binom{p}{2}$. From Figure 5.4.2, we see that for $p = 3$ the formula is

$$m_3(x, y) = 1 + x + 3x^2 + 2x^3 + y + 2xy + 3x^2y + y^2 + xy^2 + y^3. \tag{5.4.2}$$

For the derivation of the formula for $m_p(x, y)$, we use a slight modification of Pólya's Theorem in which two kinds of figure counting series are employed together with the special cycle index in two variables of the previous section. In the notation of Section 5.3, there is the identity

$$\Gamma(K_p)^{[2]} | U = S_p^{[2]}. \tag{5.4.3}$$

Figure 5.4.1

A mixed graph of order 4.

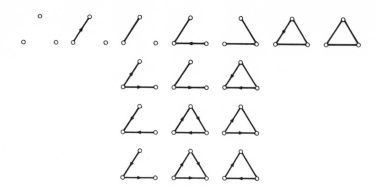

Figure 5.4.2

The 16 mixed graphs of order 3.

Thus $Z(S_p^{[2]}; s_k, t_k)$ is the cycle index of the reduced ordered pair group $S_p^{[2]}$ in which the variables s_k are used for pairs of converse cycles and t_k for self-converse cycles. We will see that on replacing each s_k in this expression by $(1 + 2x^k + y^k)^{1/2}$ and each t_k by $1 + y^{k/2}$, the polynomial $m_p(x, y)$ is obtained.

Theorem   The counting polynomial for mixed graphs of order p is given by

$$m_p(x, y) = Z(S_p^{[2]}; (1 + 2x + y)^{1/2}, 1 + y^{1/2}), \tag{5.4.4}$$

where

$$Z(S_p^{[2]}; s_k, t_k) = \frac{1}{p!} \sum_{(j)} \frac{p!}{\prod k^{j_k} j_k!} \prod_{k \text{ odd}} s_k^{(k-1)j_k}$$

$$\times \prod_{k \text{ even}} (s_k^{k-2} t_k)^{j_k} \prod_k s_k^{2k\binom{j_k}{2}} \prod_{r<w} s_{[r,w]}^{2(r,w)j_r j_w}. \tag{5.4.5}$$

As an example we give some of the details for $p = 3$. First we have the cycle index formula

$$Z(S_3^{[2]}; s_k, t_k) = (1/3!)(s_1^6 + 3s_2^2 t_2 + 2s_3^2), \tag{5.4.6}$$

Substituting the figure counting series $(1 + 2x + y)^{1/2}$ and $1 + y^{1/2}$, we obtain

$$m_3(x, y) = \tfrac{1}{6}((1 + 2x + y)^3 + 3(1 + y)(1 + 2x^2 + y^2)$$

$$+ 2(1 + 2x^3 + y^3)) \tag{5.4.7}$$

which agrees pleasantly with (5.4.2) and the mixed graphs shown in Figure 5.4.2.

Equation (5.4.5) needs no comment except to notice that it is obtained from (5.1.5) by modifying it in accordance with the part of the proof of (5.2.1) where it is shown that each even cycle of a permutation in S_p induces one self-converse cycle.

Here is a brief sketch of the proof of (5.4.4). As usual the power group $E_{2^p}^{S_p^{[2]}}$ acts on the functions in $Y^{X^{[2]}}$. Since each such function f represents a digraph with say q oriented lines and r symmetric pairs of arcs, f can also be regarded as a mixed graph with q oriented lines and r ordinary lines. Obviously, any two functions in $Y^{X^{[2]}}$ are in the same orbit of the power group if and only if their mixed graphs are isomorphic. Finally, the functions are assigned weights in the usual fashion and the weighted version (2.3.10) of Burnside's Lemma is applied to obtain (5.4.4).

The idea of $1 + 2x + y$ is that the term 1 stands for nonadjacency of the point pair, while $2x$ indicates the two possible orientations, and y an undirected line. The radical in $(1 + 2x + y)^{1/2}$ vanishes in $m_p(x, y)$ because every variable s_k occurs only with even powers since converse cycles necessarily appear in pairs. Similarly, the $1 + y$ stands as usual for no line or an ordinary line, as oriented lines simply are taboo for self-converse cycles. The radical in $1 + y^{1/2}$ also vanishes in $m_p(x, y)$ because, as shown in the only term of (5.4.5) containing t_k, k is even. This in turn holds since every self-converse cycle has even length.   //

The counting polynomials $g_p(x)$ and $d_p(x)$ for graphs and digraphs have already been derived, and the polynomial $o_p(x)$ for oriented graphs was found in [H6]. Observe that each of these polynomials is easily obtained from $m_p(x, y)$, which is thus a simultaneous generalization of three previous enumeration formulas:

$$d_p(x) = m_p(x, x^2), \qquad o_p(x) = m_p(x, 0), \qquad g_p(y) = m_p(0, y). \quad (5.4.8)$$

For $p = 3$, we find from (5.4.7) that:

$$d_3(x) = m_3(x, x^2) = 1 + x + 4x^2 + 4x^3 + 4x^4 + x^5 + x^6,$$

$$o_3(x) = m_3(x, 0) = 1 + x + 3x^2 + 2x^3,$$

$$g_3(y) = m_3(0, y) = 1 + y + y^2 + y^3. \quad (5.4.9)$$

These are quickly verified by Figure 5.4.2.

A *complete digraph* has either an oriented line or a symmetric pair of lines joining every pair of points. The digraph in Figure 5.4.3, is a complete

Figure 5.4.3

A complete digraph of order 5.

directed graph on five points with three symmetric pairs of arcs and seven oriented lines.

Let c_{pqr} be the number of complete digraphs with p points having exactly q oriented lines and r symmetric pairs. Then the polynomial $c_p(x, y)$ which enumerates complete digraphs with p points according to both the number of oriented lines and symmetric pairs is defined by

$$c_p(x, y) = \sum c_{pqr} x^q y^r \qquad (5.4.10)$$

where $q + r = \binom{p}{2}$.

From Figure 5.4.4, we see that for $p = 3$ the formula is $c_3(x, y) = 2x^3 + 3x^2 y + xy^2 + y^3$.

The enumeration formula for $c_p(x, y)$ is easily obtained by modifying the formula for mixed graphs. The integer 1 in each of the two figure counting series $(1 + 2x + y)^{1/2}$ and $1 + y^{1/2}$ represents the possibility of having no line joining a pair of points. Since in a complete digraph there is always either an oriented line or a symmetric pair joining a pair of points, the appropriate figure counting series are $(2x + y)^{1/2}$ and $y^{1/2}$.

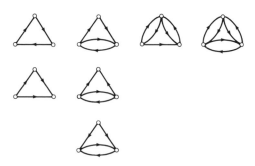

Figure 5.4.4

The complete digraphs of order 3.

Corollary The enumeration polynomial for complete digraphs on p points is given by

$$c_p(x, y) = Z(S_p^{[2]}; (2x + y)^{1/2}, y^{1/2}). \tag{5.4.11}$$

An immediate consequence of this corollary is that the number of tournaments on p points is

$$T(p) = c_p(1, 0). \tag{5.4.12}$$

Using (5.4.12) and (5.4.5), it is a matter of routine manipulation to obtain (5.2.1) and (5.2.2) explicitly.

The total number c_p of complete digraphs, regardless of the number of oriented lines or symmetric pairs, is

$$c_p = c_p(1, 1). \tag{5.4.13}$$

For example, Figure 5.4.4, shows that $c_3 = 7$.

Using the formula (5.4.5), we obtain the following expression for c_p.

Theorem The number of complete digraphs of order p is

$$c_p = \frac{1}{p!} \sum_{(j)} \frac{p!}{\prod k^{j_k} j_k!} 3^{a(j)}, \tag{5.4.14}$$

where

$$a(j) = \sum_{k=1}^{p} \left(\left[\frac{k-1}{2} \right] j_k + k \binom{j_k}{2} \right) + \sum_{r < s} (r, s) j_r j_s. \tag{5.4.15}$$

The first five values of c_p are:

p	1	2	3	4	5
c_p	1	2	7	42	582

EXERCISES

5.1 The number of complete digraphs of order p equals the number of oriented graphs of order p.

5.2 Digraphs whose points all have outdegree 2.　　　　(C. P. Lawes)

5.3 Point-symmetric digraphs of order p, a prime. (*Hint*: use the method of Turner [T1] for graphs.)

5.4   The number of tournaments of order $p \geq 5$ which admit exactly one hamiltonian cycle is

$$1 + \sum_{k=1}^{p-3} \sum_{n=0}^{m} 2^{p-k-n-4} \left[2 \binom{p-k-3}{n} \binom{k-1}{n+1} + \binom{p-k-4}{n} \binom{k-1}{n} \right]$$

where $m = \min(k-1, p-k-3)$.   (Douglas [D2])

5.5   Write an explicit formula for the cycle index $Z(S_p^2)$, and use it to get the counting polynomial for relations of order 4.

Chapter 6 | POWER GROUP ENUMERATION

If $c(x)$ is a counting series which enumerates the elements of a set Y and A is a permutation group with object set X, then we saw in Chapter 2 that Pólya's theorem provides a method for expressing the series $C(X)$, which enumerates the weighted orbits in Y^X of the power group E^A, in terms of $Z(A)$ and $c(x)$. There is a large class of problems for which it is essential to be able to enumerate orbits in Y^X of the power group B^A when B is not the identity group. A method for accomplishing this generalization of Pólya's theorem was first found by deBruijn [B5]. In this chapter we shall discuss a more natural alternative method [HP4] which simplifies the computation by eliminating superfluous differential operators and displaying explicitly the permutation group which acts on the functions, namely the power group. For this reason, we prefer to refer to deBruijn's result as the Power Group Enumeration Theorem. The applications include the enumeration of self-complementary graphs and digraphs, graphs with colored lines, finite automata, and self-converse digraphs.

135

6.1 POWER GROUP ENUMERATION THEOREM

Consider the power group B^A with object set Y^X where $X = \{x_1, \ldots, x_m\}$ and $Y = \{y_1, \ldots, y_n\}$. We begin by determining a formula for the number of orbits of B^A. To this end, we first write, using only the definition of the cycle index,

$$Z(B^A) = \frac{1}{|A| \cdot |B|} \sum_{\gamma \in B^A} \prod_{k=1}^{n^m} s_k^{j_k(\gamma)}. \tag{6.1.1}$$

For each permutation $\gamma = (\alpha; \beta)$ in B^A, the formulas for $j_k(\gamma)$ in terms of the $j_k(\alpha)$ and $j_k(\beta)$ are given by the next two equations. We first show that

$$j_1(\alpha; \beta) = \prod_{k=1}^{m} \left(\sum_{s|k} s j_s(\beta) \right)^{j_k(\alpha)}, \tag{6.1.2}$$

where $(\sum_{s|k} s j_s(\beta))^{j_k(\alpha)} = 1$ whenever $j_k(\alpha) = 0$.
For $t > 1$, we then use möbius inversion to obtain

$$j_t(\alpha; \beta) = (1/t) \sum_{s|t} \mu(t/s) j_1(\alpha^s; \beta^s). \tag{6.1.3}$$

To justify (6.1.2) and (6.1.3), consider any permutation $\gamma = (\alpha; \beta)$ in the power group B^A. Let z_k be a cycle of length k in the disjoint cycle decomposition of α. Let S be the set of k elements of X which are permuted by z_k. Then $(z_k; \beta)$ is a permutation which acts on Y^S. Define $c_k(\beta)$ as the number of functions in Y^S which are fixed by the permutation $(z_k; \beta)$. Then clearly

$$j_1(\alpha; \beta) = \prod_{k=1}^{m} (c_k(\beta))^{j_k(\alpha)}, \tag{6.1.4}$$

where $(c_k(\beta))^{j_k(\alpha)} = 1$ whenever $j_k(\alpha) = 0$.
A function f which is fixed by $(z_k; \beta)$ must assume all of its functional values in the set of elements permuted by a single cycle z_s of length s in the disjoint cycle decomposition of β. Suppose for such a function that $f(x) = y$ for some x in z_k and y in z_s. Then $(z_k; \beta)^s f(x) = \beta^s f(z_k^s x) = f(z_k^s x)$. But since f is fixed by $(z_k; \beta)$ it is also fixed by $(z_k; \beta)^s$. Therefore $(z_k; \beta)^s f(x) = f(x)$, and so $y = f(z_k^s x)$. Similarly for all i, we have $y = f(z_k^{is} x)$. Now $f(z_k x) = \beta^{-1} y$ and so the equations involving x and y also hold for $z_k x$ and $\beta^{-1} y$. We now focus on the cycle z_s containing element y and consider the part $(z_k; z_s)$ of $(z_k; \beta)$. Since a further application of function f must be compatible with the cycle lengths k and s, it is easy to see that $s|k$. Since element y in z_k is arbitrary, there are exactly s such functions for each cycle z_s. Thus $c_k(\beta) = \sum_{s|k} s j_s(\beta)$ and on substituting it in (6.1.4), we obtain (6.1.2).

Now note that if the contribution of a permutation α to $Z(A)$ is $\prod_{k=1}^{m} s_k^{j_k(\alpha)}$, then that of α^t is

$$\prod_{k=1}^{m} s_{k/(k,t)}^{(k,t)j_k(\alpha)}. \tag{6.1.5}$$

Since $(\alpha; \beta)^k = (\alpha^k; \beta^k)$, we have from (6.1.5),

$$j_1(\alpha^t; \beta^t) = \sum (k, t) j_k(\alpha; \beta) \tag{6.1.6}$$

where the sum is over all k such that $k = (k, t)$, i.e., all divisors k of t. Thus

$$j_1(\alpha^t; \beta^t) = \sum_{k|t} k j_k(\alpha; \beta) \tag{6.1.7}$$

and möbius inversion yields (6.1.3). The use of (6.1.2), (6.1.5), and (6.1.3) gives $j_t(\alpha; \beta)$ in terms of the $j_k(\alpha)$ and $j_k(\beta)$.

The next theorem is used so often that it will be convenient to refer to it by its initials, PGET. It enables us to determine the number of orbits of the power group B^A, given $Z(A)$ and $Z(B)$. The proof can be made by applying Burnside's Lemma (2.3.3) to formula (6.1.2) or as in [HP4] by using $Z(B^A)$ together with Pólya's Theorem (2.4.6). But we do not include the details here since it is the constant form of the result and is generalized in the power series form of the PGET which is developed below.

Theorem (Power Group Enumeration Theorem, constant form) The number of orbits of functions in Y^X determined by the power group B^A is

$$N(B^A) = |B|^{-1} \sum_{\beta \in B} Z(A; c_1(\beta), \ldots, c_m(\beta)) \tag{6.1.8}$$

where

$$c_k(\beta) = \sum_{s|k} s j_s(\beta). \tag{6.1.9}$$

We wish to emphasize that the constant form of the PGET amounts to an application of Burnside's Lemma to the power group.

To illustrate, we apply (6.1.8) with $A = D_4$, the dihedral group of degree 4, and $B = S_2$. The result may be interpreted in numerous ways, but perhaps the necklace context provides the most insight. The number of necklaces containing exactly four beads in which the available beads are of two *interchangeable* colors is given by the result of this substitution. From (2.4.13) we have

$$Z(D_4) = \tfrac{1}{8}(s_1^4 + 3s_2^2 + 2s_1^2 s_2 + 2s_4).$$

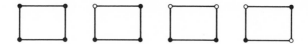

Figure 6.1.1

The four-bead necklaces with two interchangeable colors.

If β in S_2 is the identity permutation, then from (6.1.9), $c_k(\beta) = 2$ for all k. If β is the transposition, then $c_k(\beta)$ is 0 or 2 according as k is odd or even. From the PGET (6.1.8) we have

$$N(S_2^{D_4}) = \tfrac{1}{2}\{Z(D_4; 2, 2, 2, 2) + Z(D_4; 0, 2, 0, 2)\}. \tag{6.1.10}$$

We find at once that $Z(D_4; 2, 2, 2, 2) = 6$ and $Z(D_4; 0, 2, 0, 2) = 2$ and therefore from (6.1.10) the number of orbits of $S_2^{D_4}$ is 4. That is, there is one necklace in which all four beads with interchangeable colors have the same color; there is one in which exactly three beads have the same color and there are two in which exactly two beads have the same color; see Figure 6.1.1. The number of such classes of necklaces with n beads and m colors is simply the number of orbits of $S_m^{D_n}$.

6.2 SELF-COMPLEMENTARY GRAPHS

We shall now illustrate the constant form of the Power Group Enumeration Theorem by using it to determine the number of self-complementary graphs with p points. The *complement* of a graph G, denoted $\bar{G}$, has the same points as G, and two points are adjacent in $\bar{G}$ if and only if they are not adjacent in G. Then G is *self-complementary* if G and $\bar{G}$ are isomorphic. Recall from [H1, p. 24] that every self-complementary graph G has $p \equiv 0$ or 1 (mod 4) points since the number of lines in G must be $\frac{1}{2}\binom{p}{2}$, an integer. Aside from the trivial graph, the three smallest self-complementary graphs include the 4-point path and the 5-point cycle.

Read [R6] showed how to compute the number of self-complementary graphs by first applying deBruijn's theorem [B5] to count graphs modulo complementation. For this purpose we consider two graphs as *equivalent up to complementation* if they are isomorphic or one is isomorphic to the complement of the other. We now express this equivalence in terms of a power group.

Let the pair group $S_p^{(2)}$ act on $X^{(2)}$, the collection of all 2-subsets of $X = \{1, 2, \ldots, p\}$, and let S_2 have object set $Y = \{0, 1\}$. Each function f from $X^{(2)}$ to Y represents a graph G whose point set is X and in which the points i and j are adjacent whenever $f(\{i, j\}) = 1$. Thus the elements 0, 1 of Y are used to indicate the absence or presence of a line respectively. Consider the power group B^A with $A = S_p^{(2)}$ and $B = S_2$. Clearly the graphs of two functions are equivalent up to complementation if and only if the two functions are in the same orbit of this power group B^A. For if a permutation $(\alpha; (0)(1))$ in this group sends function f to g, then f and g represent isomorphic graphs. And if $(\alpha; (01))$ sends f to g, then these two functions represent complementary graphs. Hence, the number a_p of graphs up to complementation with p points is the number of orbits of this power group. Therefore we can apply the constant form of the PGET with $A = S_p^{(2)}$ and $B = S_2$. Consider the two permutations $(0)(1)$ and (01) of S_2. If $\beta = (0)(1)$, then $j_1(\beta) = 2$ and $j_s(\beta) = 0$ for $s > 1$; hence $c_k(\beta) = 2$ for all k. For $\beta = (01)$, we have $j_2(\beta) = 1$ and $j_s(\beta) = 0$ for $s \neq 2$; hence we have $c_k(\beta) = 2$ when k is even and $c_k(\beta) = 0$ when k is odd. Thus from formula (6.1.9), the number of graphs with p points up to complementation is

$$a_p = \tfrac{1}{2}\{Z(S_p^{(2)}; 2, 2, \ldots) + Z(S_p^{(2)}; 0, 2, 0, 2, \ldots)\}. \qquad (6.2.1)$$

Read then observed that $2a_p$ is the number of graphs with p points provided that a graph is counted twice if self-complementary and just once if not. It follows that the number $\bar{g}_p$ of self-complementary graphs satisfies

$$\bar{g}_p = 2a_p - g_p \qquad (6.2.2)$$

where g_p as in (4.2.1) is the number of graphs with p points. From Pólya's result (4.1.8) we know that $g_p = Z(S_p^{(2)}; 2, 2, \ldots)$. Substituting in (6.2.2) we find the formula for $\bar{g}_p$ given in the next theorem.

Theorem The number $\bar{g}_p$ of self-complementary graphs with p points is

$$\bar{g}_p = Z(S_p^{(2)}; 0, 2, 0, 2, \ldots). \qquad (6.2.3)$$

Because $p \equiv 0$ or 1 (mod 4) for $\bar{g}_p > 0$, these are the only values of p in Table 6.1.1 taken from [R6].

TABLE 6.1.1

THE NUMBER OF SELF-COMPLEMENTARY GRAPHS

p	4	5	8	9	12	13	16	17
$\bar{g}_p$	1	2	10	36	720	5 600	703 760	11 220 000

Formula (6.2.3) is specialized for these two cases as follows

$$\bar{g}_{4n} = \frac{1}{n!} \sum_{(j)} \frac{n!}{\prod k^{j_k} j_k!} 2^{c(j)},$$ (6.2.4)

and

$$\bar{g}_{4n+1} = \frac{1}{n!} \sum_{(j)} \frac{n!}{\prod k^{j_k} j_k!} 2^{c(j) + \Sigma j_k}$$ (6.2.5)

where the sums are over all partitions (j) of n and

$$c(j) = 2 \sum_{k=1}^{n} j_k(kj_k - 1) + 4 \sum_{1 \le r < t \le n} (r - t)j_r j_t.$$ (6.2.6)

The *complement* $\bar{D}$ of a digraph D has the same points as D and u is adjacent to v in $\bar{D}$ if and only if u is not adjacent to v in D. The number $\bar{d}_p$ of self-complementary digraphs may be found by the same approach as above.

Theorem   The number $\bar{d}_p$ of self-complementary digraphs with p points is

$$\bar{d}_p = Z(S_p^{[2]}; 0, 2, 0, \dots).$$ (6.2.7)

On considering the number $\bar{d}_{2n}$ of digraphs on an even number of points we find, following Read, the curious fact that

$$\bar{d}_{2n} = \bar{g}_{4n}.$$ (6.2.8)

However no one has yet found a natural 1–1 correspondence between these self-complementary graphs and digraphs. Thus $\bar{d}_4, \bar{d}_6$ and $\bar{d}_8$ can be found by using Table 6.1.1 for $p = 8, 12, 16$. These and other small values are displayed in Table 6.1.2.

TABLE 6.1.2

THE NUMBER OF SELF-COMPLEMENTARY DIGRAPHS

p	2	3	4	5	6	7	8
$\bar{d}_p$	1	4	10	136	720	44 224	703 760

Formula (6.2.7) is specialized for the odd case as follows

$$\bar{d}_{2n+1} = \frac{1}{n!} \sum_{(j)} \frac{n!}{\prod k^{j_k} j_k!} 2^{c(j) + 2\Sigma j_k}$$ (6.2.9)

where the sum is again over all partitions (j) of n.

6.3 FUNCTIONS WITH WEIGHTS

There are many enumeration problems in which integral weights are assigned to the functions so that each function in an orbit of the power group has the same weight. We seek to express the answer as the generating function

$$C(x) = C_0 + C_1 x + C_2 x^2 + \cdots \qquad (6.3.1)$$

in which C_i is the number of orbits of weight i.

Now consider the power group B^A acting on Y^X where A, X, and B are finite, but Y may be countably infinite (to allow for all nonnegative integers as weights). Let w be a function from Y into the set $\{0, 1, 2, \ldots \}$. As usual w is called a *weight* function and for each f in Y^X we define the *weight* of f, denoted $w(f)$, by

$$w(f) = \sum_{x \in X} w(f(x)). \qquad (6.3.2)$$

Since X is required to be finite, the sum in (6.3.2) is defined. If for every orbit of B^A, all functions in that orbit have the same weight, then we can define the *weight of an orbit* to be the weight of any function in it. Furthermore, if for each $i = 0, 1, 2, \ldots$, the number of elements in Y which have weight i is finite, then we can ask for the number of orbits of any specified weight. Thus we regard two functions as the "same" if and only if they are in the same orbit. Let C_k be the number of different functions of weight k determined by the power group. Then the determination of the series $C(x)$ of (6.3.1) provides the number of orbits of given weight.

We now seek to determine conditions that will ensure that functions in the same orbit of B^A have the same weight. To this end, let $Y_i = w^{-1}(i)$ be the set of all elements in Y with weight i. We have seen that each Y_i must be finite in order that there be a finite number of functions of any specified weight. By $B(Y_i)$ is meant that subset of Y consisting of all objects $\beta(y)$, $\beta \in B$, $y \in Y_i$. Now the condition required may be stated as follows.

Lemma All functions in the same orbit of the power group B^A have equal weight if and only if $B(Y_i) = Y_i$ for each $i = 0, 1, 2, \ldots$.

Proof For the sufficiency, suppose f and g are in the same orbit of B^A. Then for some $(\alpha; \beta)$ in B^A, we have $(\alpha; \beta)f = g$ so that $\beta f(\alpha x) = g(x)$ for all x in X. Note that $B(Y_i) = Y_i$ implies $w(\beta f(\alpha x)) = w(f(\alpha x))$. Therefore

$$w(g) = \sum_{x \in X} w(\beta f(\alpha x)) = \sum_{x \in X} w(f(x)) = w(f).$$

For the necessity, suppose there is an element y in Y and a permutation β in B such that $w(y) \neq w(\beta y)$. Let f and g in Y^X be defined by $f(x) = y$ and $g(x) = \beta y$ for all x in X. Then f and g are in the same orbit of B^A, but $w(f) \neq w(g)$. $//$

From now on we assume that the condition of this lemma is satisfied, namely that for all i, $B(Y_i) = Y_i$. Then each permutation β in B may be written as a product

$$\beta = \prod \beta_i \qquad (6.3.3)$$

where for each y in Y, $\beta(y) = \beta_i(y)$ if $y \in Y_i$.

To obtain the generating function $C(x)$, we now need only modify the variables $c_k(\beta)$ which appear in the statement (6.1.8) of the PGET. Let $\gamma = (\alpha; \beta)$ be any permutation in the power group B^A. Suppose z_k is any cycle of length k in the disjoint cycle decomposition of α. Again let S be the set of elements of X which are permuted by z_k. For each $i = 0, 1, 2, \ldots$, define $c_i^k(\beta)$ as the number of functions f in Y^S which are fixed by the permutation $(z_k; \beta)$ and which have

$$\sum_{x \in S} w(f(x)) = i. \qquad (6.3.4)$$

For convenience let us write the generating function

$$c_k(\beta, x) = \sum_i c_i^k(\beta)x^i. \qquad (6.3.5)$$

Note that $c_k(\beta, 1) = c_k(\beta)$ as defined in (6.1.9). Then the desired generating function $C(x)$ is given by

$$C(x) = |B|^{-1} \sum_{\beta \in B} Z(A; c_1(\beta, x), c_2(\beta, x), \ldots, c_m(\beta, x)). \qquad (6.3.6)$$

But using the same approach as made in the derivation of (6.1.2) for $j_1(\alpha; \beta)$, we have

$$c_1(\beta, x) = \sum_i j_1(\beta_i)x^i \qquad (6.3.7)$$

$$c_2(\beta, x) = \sum_i (j_1(\beta_i) + 2j_2(\beta_i))x^{2i} \qquad (6.3.8)$$

$$c_3(\beta, x) = \sum_i (j_1(\beta_i) + 3j_3(\beta_i))x^{3i}, \qquad (6.3.9)$$

and in general

$$c_k(\beta, x) = \sum_i \left(\sum_{s|k} sj_s(\beta_i) \right) x^{ki}. \qquad (6.3.10)$$

Collecting these observations, we have the following result.

Theorem (Power Group Enumeration Theorem, power series form)   The series $C(x)$ which enumerates by weight different functions as determined by the power group B^A is

$$C(x) = |B|^{-1} \sum_{\beta \in B} Z(A; c_1(\beta, x), c_2(\beta, x), \ldots, c_m(\beta, x)) \qquad (6.3.11)$$

where

$$c_k(\beta, x) = \sum_i \left(\sum_{s|k} s j_s(\beta_i) \right) x^{ki}. \qquad (6.3.12)$$

Analogous to the remark after the statement of the constant form of the PGET, this power series form consists essentially of an application of Pólya's Theorem to the cycle index of the power group.

Pólya's Theorem is immediately obtained from this theorem by taking B as the identity group on Y. Then with β the identity permutation on Y, $j_1(\beta_i) = |Y_i|$ for each i and $j_s(\beta_i) = 0$ for all $s \neq 1$. Thus $\sum_{s|k} s j_s(\beta_i) = |Y_i|$ for each i, and $c_k(\beta, x) = \sum_i |Y_i| x^{ki}$ for each $k = 1$ to m. Therefore writing $c(x^k)$ for $c_k(\beta, x)$, we have

$$C(x) = Z(A; c(x), c(x^2), \ldots, c(x^m)). \qquad (6.3.13)$$

The difference between the expression for $C(x)$ in the power series form of the PGET and that involving partial derivatives given in deBruijn's statement of the same result is merely formal, as shown in [HP4]. It is just a matter of a change in notation and routine algebraic manipulations (Exercise 6.15).

To illustrate, we return to the necklace problem discussed in Section 6.1. Let D_4 act on $X = \{1, 2, 3, 4\}$ and suppose $Y = \{1, 2, 3\}$ with $w(1) = 0$ and $w(2) = w(3) = 1$. The group B shall consist of the permutations $(1)(2)(3)$ and $(1)(23)$. Then the power series form of the PGET may be applied to obtain the series $C(x)$ and we can interpret the coefficient of x^k in this series as the number of four-beaded necklaces which have $4 - k$ beads of color 1 and k beads of the interchangeable colors 2 and 3.

In applying the theorem, first consider $\beta = (1)(2)(3)$; then $c_k(\beta, x) = 1 + 2x^k$ for each k. When $\beta = (1)(23)$ then $c_k(\beta, x) = 1$ if k is odd, and $c_k(\beta, x) = 1 + 2x^k$ if k is even. Using formula (6.3.11) in the theorem, we have

$$C(x) = \tfrac{1}{2}\{Z(D_4, 1 + x) + Z(D_4; 1, 1 + 2x^2, 1, 1 + 2x^4)\}.$$

Substitution into $Z(D_4)$ yields

$$C(x) = 1 + x + 4x^2 + 3x^3 + 4x^4.$$

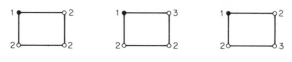

Figure 6.3.1

Necklaces with one fixed color and two interchangeable colors.

We now verify in Figure 6.3.1 that the coefficient of x^3 in the preceding equation is 3 by writing numbers 2, 3 near the points for the interchangeable colors and 1 for the fixed color.

Note that the coefficient of x^4 in the last equation is the number of four-beaded necklaces with two interchangeable colors. We have already verified in Figure 6.1.1 that this number is 4.

The PGET is readily modified to handle problems in which the weight function assumes values in any commutative ring which contains the rationals, although there does not appear to be an abundance of intuitively interesting problems at this level of generality.

6.4 GRAPHS WITH COLORED LINES

Read [R7] gives the generating function for the enumeration of graphs with p points whose lines are colored with m interchangeable colors. The power series form of the PGET provides a simple approach to the problem of determining this function.

Let A be the pair group $S_p^{(2)}$ with object set $X^{(2)}$. Let $Y = \{0, 1, \ldots, m\}$ and consider the symmetric group S_{m+1} acting on Y. For B we take the subgroup of S_{m+1} which fixes the element 0 of Y so that $Z(B) = s_1 Z(S_m)$. Next we define a weight function w from Y into the set $\{0, 1\}$ with $w(y) = 0$ if and only if $y = 0$. Then each function f from $X^{(2)}$ to Y represents a graph with $f^{-1}(i)$ lines of "color i" for $i = 1$ to m. Further, the weight $W(f)$, as defined in (6.3.2) is the number of lines in the graph represented by f. It follows that the generating function $N_p^m(x)$ which we seek is obtained by applying (6.3.11) to this power group B^A.

To illustrate, we show some of the details for $m = 3$. In accordance with the notation above, we have $Y_0 = \{0\}$ and $Y_1 = \{1, 2, 3\}$. For each β in B we must compute $c_k(\beta, x)$ as given by formula (6.3.10). Recall that for $i = 0, 1$, the coefficient of x^{ki} in $c_k(\beta, x)$ is $\sum_{s|k} s j_s(\beta_i)$. There are three cases, one for each type of permutation.

Case i $\beta = (0)(1)(2)(3)$.

We have $\beta_0 = (0)$ and $\beta_1 = (1)(2)(3)$. Therefore $j_1(\beta_0) = 1$ and $j_1(\beta_1) = 3$. So $c_k(\beta, x) = 1 + 3x^k$ for all k.

Case ii $\beta = (0)(12)(3)$

Since $\beta_0 = (0)$ and $\beta_1 = (12)(3)$, we have $j_1(\beta_0) = 1$, $j_1(\beta_1) = 1$, and $j_2(\beta_1) = 1$. Therefore $\sum_{s|k} s j_s(\beta_1)$ is $j_1(\beta_1)$ or $(j_1(\beta_1) + 2j_2(\beta_1))$ according as k is odd or even. Hence $c_k(\beta, x)$ is $1 + x^k$ or $1 + 3x^k$ according as k is odd or even.

Case iii $\beta = (0)(123)$

Since $\beta_0 = (0)$ and $\beta_1 = (123)$ we have $j_1(\beta_0) = 1$ and $j_3(\beta_1) = 1$. Therefore $\sum_{s|k} s j_s(\beta_1)$ is 0 or $3j_3(\beta_1)$ according as $3 \nmid k$ or $3|k$. Hence $c_k(\beta, x)$ is 1 or $1 + 3x^k$ according as $3 \nmid k$ or $3|k$.

If $\beta = (0)(13)(2)$ or $\beta = (0)(23)(1)$, then, of course, $c_k(\beta, x)$ is given by Case ii. From (6.3.11) we now have

$$N_p^3(x) = \tfrac{1}{6}(Z(S_p^{(2)}; 1 + 3x, 1 + 3x^2, \ldots)$$
$$+ 3Z(S_p^{(2)}; 1 + x, 1 + 3x^2, 1 + x^3, \ldots)$$
$$+ 2Z(S_p^{(2)}; 1, 1, 1 + 3x^3, \ldots)).$$

Therefore, for $p = 3$,

$$N_3^3(x) = 1 + x + 2x^2 + 3x^3.$$

In Figure 6.4.1 we illustrate this equation by showing all the 3-point graphs in which the lines are assigned three interchangeable colors a, b, c.

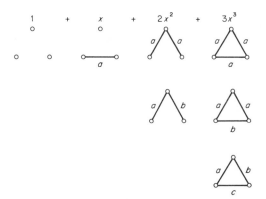

Figure 6.4.1

Graphs with lines of interchangeable colors.

6.5 FINITE AUTOMATA

Harrison [H15] first solved the problem of finding the number of different automata. With the aid of the PGET, the problem of counting automata with specified initial and final states can also be handled routinely, as in [HP9]. To set the stage, we enumerate ordered pairs of functions with respect to the product of two power groups. Finite automata are then concisely defined as certain ordered pairs of functions. We review the enumeration of automata in the natural setting of the power group, and then extend this result to provide for initial and terminal states.

To develop the enumeration theorem for ordered pairs of functions, let B_1 and B_2 be permutation groups. Let $A_1 = C_1 \times D_1$ and $A_2 = C_2 \times D_2$ be two products of groups where for $i = 1, 2$ the degrees of C_i and D_i are c_i and d_i respectively. Using formula (4.3.10) to get the cycle index of the cartesian product of two groups, equation (6.1.2) for the number of objects fixed by a given permutation in the power group, and Burnside's Lemma (2.3.3), we can obtain the next theorem.

Theorem The number of orbits determined by any subgroup F of $(B_1^{C_1 \times D_1}) \times (B_2^{C_2 \times D_2})$ is

$$N(F) = |F|^{-1} \sum{}^* \prod_{i=1}^{2} \left\{ \prod_{p=1}^{c_i} \prod_{q=1}^{d_i} \left[\sum_{s \mid [p,r]} s j_s(\beta_i) \right]^{j_p(\gamma_i) j_q(\delta_i)(p,q)} \right\} \qquad (6.5.1)$$

where the asterisk indicates the sum is taken over all permutations $(((\gamma_1, \delta_1); \beta_1), ((\gamma_2, \delta_2); \beta_2))$ in F.

There are a number of ways in which finite automata can be defined, but the formulation most convenient for our purposes may be expressed in terms of ordered pairs of functions. Let X, Y, and S be three sets with cardinalities k, m, and n respectively. The elements of S will be called *states*; the sets X and Y, the *input* and *output alphabets*, respectively. An *automaton* is an ordered pair of functions (f_1, f_2) with $f_1 : S \times X \to S$ and $f_2 : S \times X \to Y$. The map f_1 is called the *input function* and f_2 is the *output function*. In conventional terminology, f_1 tells the next state and f_2 the output symbol when the automaton is in any given state and is presented with some input symbol.

Three types of equivalence for automata are described by Harrison [H15], but we will treat just one of these types here; namely isomorphism; the others may be handled similarly. Let S_k, S_m, and S_n be the symmetric groups with object sets X, Y, and S respectively. Thus there are n states, k input letters, and m output letters. Two automata (f_1, f_2) and (g_1, g_2) are called *isomorphic* if there are permutations α in S_n, β in S_k, and γ in S_m such that for all s in S and x in X

$$f_1(s, x) = \alpha^{-1}g_1(\alpha s, \beta x) \tag{6.5.2}$$

and

$$f_2(s, x) = \gamma^{-1}g_2(\alpha s, \beta x). \tag{6.5.3}$$

Thus equation (6.5.2) allows for changing the names of the states, while (6.5.3) admits permuting the input symbols. Figure 6.5.1 shows two automata which represent the same input function under the equivalence relation of isomorphism defined above. The symbols 0 and 1 are used for the input alphabet. Both the two state labels and the two input letters have been interchanged.

In order to have an appropriate graph theoretic setting, we require the next concept. In a *net*, both loops and multiple directed lines are permitted; see [HNC1, p. 5]. If the outdegree of every point is k, and each of the k lines from a point is given a different label from the input alphabet X, then such a net represents the input function of an automaton. We also label the points of the net as the states of the automaton at hand.

To further clarify the definition of isomorphic automata given above, consider equation (6.5.2) which defines equivalence for input functions f_1 and g_1. In the labeled net of f_1, there is a directed line with input x from each state s to the state $f_1(s, x)$. Similarly, in the net of g_1 there is a directed line with input label βx from each state αs to the state $g_1(\alpha s, \beta x)$. Thus the permutation $((\alpha, \beta); \alpha^{-1})$ in the power group $S_n^{S_n \times S_k}$ sends the input function g_1 to f_1 and simply changes the names of the states along with the appropriate changes in the input labels on the directed lines.

For the enumeration of automata, we must take

$$C_1 = C_2 = S_n, \qquad D_1 = D_2 = S_k, \qquad B_1 = S_n, \qquad \text{and} \qquad B_2 = S_m.$$

Then we apply the constant form of the PGET with F the subgroup of $(S_n^{S_n \times S_k}) \times (S_m^{S_n \times S_k})$ which consists of all permutations of the form $(((\alpha, \beta); \alpha^{-1}, ((\alpha, \beta); \gamma))$. Since the order of F is $n!k!m!$, the formula is as follows.

Figure 6.5.1

Two isomorphic automata.

Corollary   The number $a(n, k, m)$ of automata with n states, k input symbols, and m output symbols is

$$a(n, k, m) = (1/n!k!m!) \sum I(\alpha, \beta, \alpha)I(\alpha, \beta, \gamma), \qquad (6.5.4)$$

where the sum is over all permutations in F and

$$I(\alpha, \beta, \gamma) = \prod_{p=1}^{n} \prod_{q=1}^{k} \left(\sum_{s|[p,q]} sj_s(\gamma) \right)^{j_p(\alpha)j_q(\beta)(p,q)}. \qquad (6.5.5)$$

Obviously (6.5.4) can be modified further by using the formula for the number of permutations in a symmetric group with a given partition.

As an illustration we give some of the details for finding $a(2, 2, 1)$, the number of automata with two states, two input symbols, and just one output symbol. Since there is only one output function, formula (6.5.4) is somewhat simplified:

$$a(2, 2, 1) = \tfrac{1}{4}\sum \prod_{p=1}^{2} \prod_{q=1}^{2} \left(\sum_{s|[p,q]} sj_s(\alpha) \right)^{j_p(\alpha)j_q(\beta)(p,q)}$$

$$= \tfrac{1}{4}(2^4 + 2^2 + 2^2 + 2^2) = 7. \qquad (6.5.6)$$

In an automaton one usually distinguishes one of the states, calling it the *initial state* or *source*. Further, one may distinguish several other states

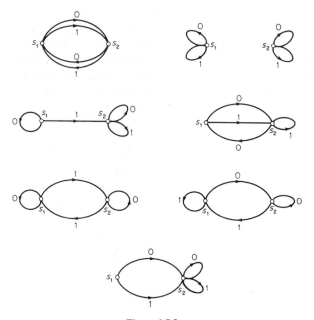

Figure 6.5.2

The seven automata of equation (6.5.6).

called *terminal states*. Thus to enumerate these automata, we must enumerate appropriately rooted nets. More specifically, we enumerate nets with one initial state and t terminal states. The method involves application of the power group to the original enumeration of rooted graphs given in [H4].

The operations on permutation groups of forming the power group, product and sum provide the means for an explicit description of the permutation group which accomplishes the enumeration. Let H be the permutation group

$$H = \{(E_1 S_{n-t-1} S_t)^{(E_1 S_{n-t-1} S_t) \times S_k}\} \times \{S_m^{(E_1 S_{n-t-1} S_t) \times S_k}\}$$

acting on $S^{S \times X} \times Y^{S \times X}$.

Let $a(n, k, m, t)$ be the number of automata with n states, including one initial state and t terminal states, k input symbols, and m output symbols. Let F be the subgroup of H which consists of all permutations of the previously encountered form $(((\alpha, \beta); \alpha^{-1}), ((\alpha, \beta); \gamma))$. Then the order of F is $(n - t - 1)! t! k! m!$. As before, the number of such automata is given by formula (6.5.1).

Corollary The number of automata with one initial state and t terminal states is

$$a(n, k, m, t) = [1/(n - t - 1)! t! k! m!] \sum I(\alpha, \beta, \alpha) I(\alpha, \beta, \gamma) \qquad (6.5.7)$$

where the sum is over all permutations in H of the form $(((\alpha, \beta); \alpha^{-1}), (\alpha, \beta); \gamma))$, and $I(\alpha, \beta, \gamma)$ is given by (6.5.5).

For a simple example, we take the case in which the number m of output symbols and the number t of terminal states are both 1, and the number of input symbols is 2. Then we have

$$a(n, 2, 1, 1) = [1/(n - 2)! 2] \sum \prod_{p=1}^{n-2} \prod_{q=1}^{2} \left(\sum_{s|[p,q]} s j_s(\alpha) \right)^{j_p(\alpha) j_q(\beta)(p,q)}. \qquad (6.5.8)$$

It is now easy to calculate that for $n = 2$, $a(2, 2, 1, 1) = 10$. (See Table 6.5.1 and Figure 6.5.3.) In Table 6.5.1 the values of $a(n, 2, 1, t)$ are shown for

TABLE 6.5.1

THE NUMBERS $a(n, 2, 1, t)$

n \ t	1	2	3	4
2	10			
3	378	198		
4	16 576	16 576	5 614	
5	819 420	1 226 900	819 420	206 495

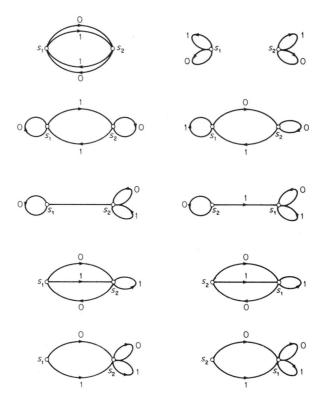

Figure 6.5.3

The ten automata verifying the value of a(2, 2, 1, 1).

small n and $t = 1$ to $n - 1$. The identical entries occur because $a(n, 2, 1, t) = a(n, 2, 1, n - t - 1)$ for $t = 1$ to $n - 2$. We note that the enumeration given by the Corollary (6.5.7) entails t terminal states different from the initial state. To admit the situation where the initial state is itself one of the terminal states, one replaces each occurrence of t in (6.5.7) by $t - 1$. It is just as easy to count automata with any number r of initial states and t terminal states, as well as any specified number of states which are both initial and terminal.

6.6 SELF-CONVERSE DIGRAPHS

Our object is to derive a formula for the counting polynomial $d'_p(x)$ which has as the coefficient of x^q, the number of *self-converse* digraphs with p points and q lines [HP5]. Such a digraph D has the property that its *converse*

digraph D' obtained from *D* by reversing the orientation of all lines is isomorphic to *D*. The derivation uses Pólya's Theorem as applied to the restriction of the power group wherein the permutations act only on 1–1 functions,
for reasons explained below. By inspection, we find that the counting polynomial $d'_3(x)$ which enumerates the self-converse digraphs with three points is

$$d'_3(x) = 1 + x + 2x^2 + 2x^3 + 2x^4 + x^5 + x^6. \tag{6.6.1}$$

Recall that the *complement* $\bar{D}$ of a digraph *D* has the same set of points as
D, and in it *u* is adjacent to *v* if and only if *u* is not adjacent to *v* in *D*. It is
easy to see that $(\bar{D})' = \overline{D'}$ i.e., the converse and the complement of a digraph
commute. This remark accounts for the end-symmetry of the coefficients of
$d'_p(x)$ as in (6.6.1) for $p = 3$.

Two digraphs D_1 and D_2 with the same set of points are *equivalent up to
conversion* if either $D_1 \cong D_2$ or $D'_1 \cong D_2$. (Note the analogy to equivalence
of graphs up to complementation.) Our objective now is to find a formula for
$c_p(x)$, the counting polynomial which enumerates digraphs with *p* points
up to conversion. To do this, we must find, as in the case for graphs, the
appropriate permutation group to which PET may be applied.

Let S_2 act on $\{1, 2\}$ and consider the power group $S_p^{S_2}$ acting on $X^{\{1,2\}}$,
the functions from $\{1, 2\}$ into *X*. Observe the natural correspondence between
the elements of $X^{[2]}$ (which are ordered pairs of distinct elements in *X*) and the
1–1 functions in $X^{\{1,2\}}$. Each ordered pair (i, j) in $X^{[2]}$ corresponds to the 1–1
function in $X^{\{1,2\}}$ which sends 1 to *i* and 2 to *j*. Thus we may consider the
restricted power group $S_p^{S_2*}$ where the restriction is to 1–1 functions as acting
on the elements of $X^{[2]}$. More specifically, the permutations of $S_p^{S_2*}$ consist
of those ordered pairs $(\alpha; \beta)$ of permutations α in S_2, β in S_p such that for any
(i, j) in $X^{[2]}$,

$$(\alpha; \beta)(i, j) = \begin{cases} (\beta i, \beta j) & \text{if} \quad \alpha = (1)(2) \\ \\ (\beta j, \beta i) & \text{if} \quad \alpha = (12). \end{cases} \tag{6.6.2}$$

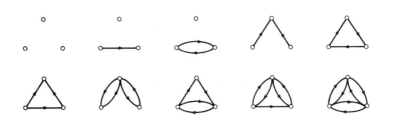

Figure 6.6.1

The ten self-converse digraphs on three points.

Now let E_2 be the identity group acting on the set $Y = \{0, 1\}$. Consider the power group B consisting of E_2 raised to the group $S_p^{S_2^*}$ acting on $Y^{X^{[2]}}$ the functions from $X^{[2]}$ into Y. Each such function f represents a digraph whose points are the elements of $X = \{1, 2, \ldots, p\}$, where i is adjacent to j whenever $f(i, j) = 1$. Thus the elements 0 and 1 of Y indicate the absence or presence of arcs.

Let f_1 and f_1 be two of these functions, and let their digraphs be D_1 and D_2 respectively. Then $D_1' \cong D_2$ or $D_1 \cong D_2$ if and only if there is a permutation γ in B such that $\gamma f_1 = f_2$. This follows from the fact that for $\gamma = ((\alpha; \beta);$ $\epsilon)$ the digraph of γf_1 is isomorphic to D_1 or D_1' according as α is (1)(2) or (12).

Thus equivalence of digraphs up to conversion corresponds to the equivalence of functions in $Y^{X^{[2]}}$ determined by the power group B.

Now by applying PET, which we can do because the base group is E_2, we obtain the next result, which is required for counting self-converse digraphs.

Theorem The counting polynomial $c_p(x)$ which enumerates digraphs up to conversion is

$$c_p(x) = Z(S_p^{S_2^*}, 1 + x). \tag{6.6.3}$$

There are formulas which can be used to express the cycle index of any restricted power group B^{A^*}, where the restriction is to the 1–1 functions which are present. But in the special case $A = S_2$ and $B = S_p$, a more explicit formula can be given. For each permutation α in S_p, the partition of α is denoted as previously by $(j) = (j_1, j_2, \ldots, j_p)$, where j_k is the number of disjoint cycles of length k in α. Then the contribution to $Z(S_p^{S_2^*})$ of $((12); \alpha)$ is

$$
\begin{aligned}
I(\alpha) = \prod_{k=1}^{p} s_{[2,k]}^{(2,k)k\binom{j_k}{2}} \cdot \prod_{\substack{1 \le r \le t \le p \\ k=[r,t]}} s_{[2,k]}^{(2,k)(r,t)j_r j_t} \\
\cdot \prod_{k \text{ odd}} s_{2k}^{[(k-1)/2]j_k} \prod_{k \text{ even}} s_k^{(k-2)j_k} s_{k/2}^{\eta(k)2j_k} s_k^{(1-\eta(k))j_k},
\end{aligned}
\tag{6.6.4}
$$

where $\eta(k) = 1$ if $k/2$ is an odd integer and 0 otherwise.

Hence the cycle index of $S_p^{S_2^*}$ can be expressed as

$$Z(S_p^{S_2^*}) = (1/2p!)\left(p! Z(S_p^{[2]}) + \sum_{\alpha \in S_p} I(\alpha)\right). \tag{6.6.5}$$

Now we make the simple observation for self-converse digraphs which corresponds to that made by Read for self-complementary graphs. Namely, the polynomial $2c_p(x)$ counts each digraph twice if it is self-converse and once if not. Hence the polynomial $2c_p(x) - d_p(x)$ counts each self-converse digraph just once. Thus we have

$$d'_p(x) = 2c_p(x) - d_p(x).$$ (6.6.6)

This together with formulas (5.1.4) and (6.6.3) gives the next result.

Theorem The counting polynomial $d'_p(x)$ for self-converse digraphs is

$$d'_p(x) = 2Z(S_p^{S_2*}, 1 + x) - Z(S_p^{[2]}, 1 + x).$$ (6.6.7)

To use formula (6.6.7) for $d'_p(x)$ let

$$F(S_p^{S_2*}) = (1/p!) \sum_{\alpha \in S_p} I(\alpha).$$

Even though this last expression does not refer to a permutation group, we still define the substitution of $1 + x$ in it to mean the polynomial obtained by replacing each variable s_k in $F(S_p^{S_2*})$ by $1 + x^k$. Combining (6.6.7) and formula (6.6.5) for $Z(S_p^{S_2*})$, we obtain

$$d'_p(x) = F(S_p^{S_2*}, 1 + x).$$ (6.6.8)

To illustrate, we develop the polynomial $d'_3(x)$ for the self-converse digraphs on three points. The cycle index of the symmetric group S_3 is

$$Z(S_3) = \tfrac{1}{6}(s_1^3 + 3s_1 s_2^2 + 2s_3).$$ (6.6.9)

From this and formula (6.6.4) for $I(\alpha)$, we have

$$F(S_3^{S_2*}) = \tfrac{1}{6}(s_2^3 + 3s_1^2 s_2^2 + 2s_6).$$ (6.6.10)

Formula (6.6.8) gives for $p = 3$,

$$d'_3(x) = \tfrac{1}{6}((1 + x^2)^3 + 3(1 + x)^2(1 + x^2)^2 + 2(1 + x^6))$$
$$= 1 + x + 2x^2 + 2x^3 + 2x^4 + x^5 + x^6.$$ (6.6.11)

and for $p = 4$,

$$d'_4(x) = \tfrac{1}{24}((1 + x^2)^6 + 6(1 + x)^2(1 + x^2)^5 + 8(1 + x^6)^2$$
$$+ 3(1 + x)^4(1 + x^2)^4 + 6(1 + x^4)^3)$$
$$= 1 + x + 3x^2 + 5x^3 + 9x^4 + 10x^5 + 12x^6$$
$$+ 10x^7 + 9x^8 + 5x^9 + 3x^{10} + x^{11} + x^{12}.$$ (6.6.12)

Figure 6.6.2

Five self-converse digraphs.

These coefficients may be checked in detail by examining the diagrams of the four point digraphs in [HP6] or in the Appendix of [H1]. In Figure 6.6.2 we show the five self-converse digraphs with four points and three lines.

A slight modification of formula (6.6.7) results in the polynomial $r'_p(x)$ that enumerates self-converse digraphs in which loops are permitted. On permitting the addition of loops to digraphs, we obtain, of course, just relations. It is easy to see how the power group $S_p^{S_2}$ can be used to count relations up to conversion. Recall that the ordered pair group S_p^2 acts on all ordered pairs (where the elements need not be distinct) as induced by the symmetric group S_p. As shown in (5.1.13), the polynomial $r_p(x)$ which counts relations is

$$r_p(x) = Z(S_p^2, 1 + x). \tag{6.6.13}$$

Then $r'_p(x)$ is given by

$$r'_p(x) = 2Z(S_p^{S_2}, 1 + x) - Z(S_p^2, 1 + x). \tag{6.6.14}$$

To use equation (6.6.14), it is convenient to introduce the following notation for each permutation α in S_p

$$J(\alpha) = I(\alpha) \sum_{k=1}^{p} s_k^{jk}. \tag{6.6.15}$$

Then the cycle index of the power group $S_p^{S_2}$ can be expressed:

$$Z(S_p^{S_2}) = \frac{1}{2p!}\left(p!\,Z(S_p^2) + \sum_{\alpha \in S_p} J(\alpha)\right). \tag{6.6.16}$$

Now let

$$H(S_p^{S_2}) = \frac{1}{p!} \sum_{\alpha \in S_p} J(\alpha). \tag{6.6.17}$$

Then the formula for $r'_p(x)$ can be written

$$r'_p(x) = H(S_p^{S_2}, 1 + x). \tag{6.6.18}$$

Let d'_p be the total number of self-converse digraphs with p points. Then, referring to (6.6.7), we see that $d'_p = d'_p(1)$. In order to express a formula

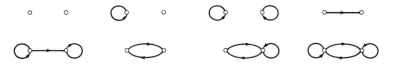

Figure 6.6.3

The self-converse relation on two points.

for d'_p in relatively manageable form, we introduce the following notation. For each α in S_p, let

$$\varepsilon(\alpha) = \sum_{k=1}^{p} \left[(2, k) \left\{ \frac{k-1}{2} j_k + k \binom{j_k}{2} \right\} + \eta(k) j_k \right]$$

$$+ \sum_{1 \le r \le t \le p} (2, [r, t])(r, t) j_r j_t. \tag{6.6.19}$$

Since the replacement in (6.6.8) and (6.6.4) of each s_k in $F(S_p^{S_2^*})$ by 2 gives $d'_p(1)$, we have

$$d'_p = (1/p!) \sum_{\alpha \in S_p} 2^{\varepsilon(\alpha)}. \tag{6.6.20}$$

A similar formula is easily obtained for the total number r'_p of self-converse relations with p points. The total for $p = 1$ to 6 are in Table 6.6.1. The eight self-converse relations on two points are drawn in Figure 6.6.3.

TABLE 6.6.1

THE NUMBER OF SELF-CONVERSE DIGRAPHS
AND RELATIONS ON p POINTS

p	1	2	3	4	5	6
d'_p	1	3	10	70	708	15 248
r'_p	2	8	44	436	7 176	222 368

EXERCISES

6.1 (a) How many orbits of functions from a set of m elements to a set of n elements are determined by the power groups $E_n^{E_m}$, $E_n^{S_m}$, and $S_n^{E_m}$?
(b) How many *onto* functions are determined by these power groups?
(*Hint:* Use binomial coefficients or Stirling numbers of the second kind.)
(Palmer [P2])

6.2   Find formulas for $Z(S_3^{C_3})$, $Z(C_3^{S_3})$, $Z(S_3^{C_3^*})$, and $Z(C_3^{S_3^*})$.   (Read [R6])

6.3 $\bar{g}_p = \sum_q (-1)^q g_{p,q}$ (Frucht and Harary [FH1])

6.4   Find the series $C(x, y)$ such that the coefficient of $x^r y^s$ is the number of six-beaded necklaces which have r beads of two interchangeable colors, s beads of two other interchangeable colors and $6 - r - s$ red beads.

6.5   Find the series $N(x, y)$ which has as the coefficient of $x^r y^s$ the number of graphs with four points and r lines of two interchangeable colors and s lines of three other interchangeable colors.

6.6   In how many ways can the faces of a cube be colored with three colors, two of which are interchangeable?

6.7   In how many ways can the faces of the cube be colored with two interchangeable colors?

6.8   Find a formula for $Z(S_2^{S_n})$.

6.9   Self-complementary eulerian graphs.   (Robinson [R18])

6.10   Self-converse oriented graphs, and self-complementary oriented graphs.   (Sridharan [S5])

6.11   The number of self-complementary tournaments of order $2n$ is

$$\frac{1}{(2n)!} \sum_{(j)}^* \frac{(2n)!}{\prod k^{j_k} j_k!} \prod_k 2^{k j_k^2/2} \prod_{r<t} 2^{(r,t) j_r j_t}$$

where the asterisk indicates the summation is over all partitions (j) of $2n$ with $j_k \neq 0$ only when k is even but $4 \nmid k$. Self-complementary tournaments of odd order.

6.12   *Vacuously transitive* relations, and digraphs in which there are no transitive triples.   (Sharp [S2])

6.13   Digraphs whose converse and complement are isomorphic.

                                                       (Palmer [P6])

6.14   A signed graph is *self-negational* if it is isomorphic to the signed graph obtained by changing positive lines to negative and vice versa. Self-negational signed graphs.   (Read [R7])

6.15   In order to state deBruijn's polynomial form [B5] of the PGET it is convenient to assume that $B = \prod_{i=0}^r B_i$ with B_i acting on Y_i. In many applications B has this form. Let

$$Z(B_i) = \frac{1}{|B_i|} \sum_{\beta \in B_i} \prod_k b_{i,k}^{j_k(\beta)}.$$

Then the polynomial $C(x)$ which counts orbits of the power group B^A is obtained by first setting

$$b_{i,k} = \exp \left\{ k \sum_{s=1}^{[m/k]} z_{ks} x^{iks} \right\}$$

in $Z(B)$. Then

$$C(x) = \left[Z\left(A; \frac{\partial}{\partial z_1}, \ldots, \frac{\partial}{\partial z_m} \right) Z(B) \right]_{z_k = 0} .$$

If it tastes good already, it will taste even better with paprika on it.

Old Hungarian saying

If an important decision is to be made, they discuss the question when they are drunk, and the following day the master of the house where the discussion was held submits their decision for reconsideration when they are sober. If they still approve it, it is adopted; if not, it is abandoned. Conversely, any decision they make when they are sober, is reconsidered afterwards when they are drunk.

Herodotus on the Persians

Chapter 7 | SUPERPOSITION

There are several natural ways to define the union of a collection of graphs. In each case we shall call the union a "superposition." Three of these are illustrated in Figure 7.0.1. The problem of determining the number of "different" superpositions which can be obtained by superposing a given set of graphs has been solved for several interesting special cases. Redfield [R10] was the first to obtain such a solution. His enumeration theorem, when

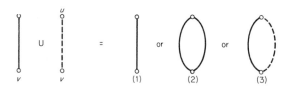

Figure 7.0.1

Three different superpositions.

combined with his "decomposition theorem," also lends itself to the enumeration of graphs and digraphs. The object of this chapter is to present Redfield's enumeration methods while recasting them in contemporary notation and terminology (see [HP8], [P4], [F2]).

7.1 REDFIELD'S ENUMERATION THEOREM

The operations cap $\cap$ and cup $\cup$ on cycle indexes were used by Redfield to express the results in both his "enumeration theorem" and his "decomposition theorem." They were exploited by Read to derive his "superposition theorem" [R1], and were interpreted by Foulkes [F1] as scalar products of certain group characters. These two operations can be introduced in a very general setting.

The ring of rational polynomials in the variables $s_1 s_2, \ldots, s_d$ is denoted by R. The operation *cap* is first defined for a sequence $s_1^{i_1} s_2^{i_2} \cdots s_d^{i_d}, s_1^{j_1} s_2^{j_2} \cdots s_d^{j_d}, \ldots$ of $m \geq 2$ monomials in R by

$$(s_1^{i_1} s_2^{i_2} \cdots s_d^{i_d}) \cap (s_1^{j_1} s_2^{j_2} \cdots s_d^{j_d}) \cap \cdots = \left(\prod_{k=1}^{d} k^{i_k} i_k! \right)^{m-1} \tag{7.1.1}$$

if $i_k = j_k = \cdots$ for all k, and is 0 otherwise ($b^0 = 1$ even when $b = 0$).

By linearity, the cap operation may then be extended to arbitrary polynomials in these variables.

The second operation, *cup*, is defined for monomials in terms of cap:

$$(s_1^{i_1} s_2^{i_2} \cdots s_d^{i_d}) \cup (s_1^{j_1} s_2^{j_2} \cdots s_d^{j_d}) \cup \cdots$$

$$= [(s_1^{i_1} s_2^{i_2} \cdots s_d^{i_d}) \cap (s_1^{j_1} s_2^{j_2} \cdots s_d^{j_d}) \cap \cdots] s_1^{i_1} \cdots s_d^{i_d}. \tag{7.1.2}$$

This operation is also extended linearly so that it also is defined for cup products of m polynomials.

Thus the result of applying the cap operation is a number whereas the cup leads to a polynomial. In practice, these operations are applied only to the cycle indexes of m permutation groups of the same degree. To illustrate, we give the details for a case where $m = 2$ with the cyclic and dihedral groups of degree 4:

$$Z(C_4) = \tfrac{1}{4}(s_1^4 + s_2^2 + 2s_4) \tag{7.1.3}$$

$$Z(D_4) = \tfrac{1}{8}(s_1^4 + 2s_1^2 s_2 + 3s_2^2 + 2s_4) \tag{7.1.4}$$

$$Z(C_4) \cap Z(D_4) = \tfrac{1}{32}[(s_1^4 \cap s_1^4) + 3(s_2^2 \cap s_2^2) \tag{7.1.5}$$

$$+ 4(s_4 \cap s_4)] = \tfrac{1}{32}(24 + 24 + 16) = 2$$

$$Z(C_4) \cup Z(D_4) = \tfrac{1}{32}(24 s_1^4 + 24 s_2^2 + 16 s_4). \tag{7.1.6}$$

Thus $Z(C_4) \cap Z(D_4)$ is just the coefficient sum of $Z(C_4) \cup Z(D_4)$.

Let W be the collection of $m \times n$ matrices in which the elements of each row are the n objects in a set S. Thus there are $(n!)^m$ matrices in W. Two matrices in W are said to be *column equivalent* if one can be obtained from the other by a permutation of the columns, and it is easy to see that there are $(n!)^{m-1}$ corresponding equivalence classes. Next, another equivalence relation is defined for the column-equivalence classes. Let C_1 and C_2 be two such classes and let $B_1, \ldots, B_m$ be permutation groups with object set S. Then C_1 and C_2 are called *equivalent with respect to* $(B_1, \ldots, B_m)$ if there is a sequence $\beta_1, \ldots, \beta_m$ of permutations with β_i in B_i for each i and a matrix $[w_{ij}]$ in C_1 such that $[\beta_i w_{ij}]$ is in C_2. That is, β_i permutes the elements in the ith row of some matrix in C_1 and the result is a matrix in C_2. Redfield's enumeration theorem expresses the number of these classes in terms of the cycle indexes of the groups B_i and the cap operation.

Redfield's Enumeration Theorem   The number $N[B_1, \ldots, B_n]$ of classes of $m \times n$ matrices equivalent with respect to the permutation groups $(B_1, \ldots, B_n)$ is

$$N[B_1, \ldots, B_m] = Z(B_1) \cap \cdots \cap Z(B_m). \tag{7.1.7}$$

This result can be verified by first constructing a permutation group that has as its orbits the classes of matrices to be counted. Then (7.1.7) follows from Burnside's Lemma (2.3.3) applied to this group.

We now show how the theorem can be used to calculate the number of superpositions of a set of graphs. Let $G_1, \ldots, G_m$ be m graphs each of which has the same set of n points and in which, for each $i = 1$ to m, the lines of G_i are labeled with the integer i (or belong to color class i). A *superposition* of these graphs has the same set of points and any two of them, say u and v are adjacent with line label i whenever u and v are adjacent in G_i. Thus the lines in a superposition are labeled but the points are not. To illustrate we display in Figure 7.1.1 all the superpositions composed of two cycles of order 5. We use solid and dashed lines to indicate the two colors.

Redfield [R10] and Read [R1] observed that the number of different superpositions is simply the number of classes of $m \times n$ matrices which are

Figure 7.1.1

The superpositions of two cycles.

equivalent with respect to $(\Gamma(G_1), \ldots, \Gamma(G_m))$. Thus we can determine the number of superpositions, provided we know the cycle indexes $Z(G_i)$ of the groups $\Gamma(G_i)$ of the graphs involved. This writing of $Z(G)$ in place of $Z(\Gamma(G))$ is a convenient abuse of notation.

Corollary The number of different superpositions of m graphs G_i with the same set of unlabeled points is

$$Z(G_1) \cap \cdots \cap Z(G_m). \tag{7.1.8}$$

For example, to determine the number of superpositions of two cycles of order n we can calculate $Z(D_n) \cap Z(D_n)$, since the group of a cycle is D_n. With $n = 5$ we know from (2.2.11) that

$$Z(D_5) = \tfrac{1}{10}(s_1^5 + 4s_5 + 5s_1 s_2^2). \tag{7.1.9}$$

From the definition of the cap operation it follows that

$$Z(D_5) \cap Z(D_5) = \tfrac{1}{100}(s_1^5 \cap s_1^5 + 16s_5 \cap s_5 + 25s_1 s_2^2 \cap s_1 s_2^2)$$

$$= \tfrac{1}{100}(120 + 80 + 200) = 4, \tag{7.1.10}$$

and this is verified by the superpositions of Figure 7.1.1.

Redfield's Enumeration Theorem can be used to determine the number of superpositions when the constituents are directed graphs or both graphs and digraphs. Indeed, Redfield illustrated his theorem by superposing cycles and directed cycles separately and together. Figure 7.1.2 shows the two superpositions of a directed and an undirected cycle of order 4, thus verifying equation (7.1.5).

Finally, suppose G is a graph with n points and just one line. Then the number of superpositions of m copies of G is equal to the number of multigraphs with n points and in which each line has a different color. Since $Z(G) = Z(S_2)Z(S_{n-2})$, the number of these is the $\cap$-product of length m:

$$Z(S_2)Z(S_{n-2}) \cap \cdots \cap Z(S_2)Z(S_{n-2}).$$

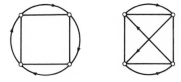

Figure 7.1.2

Two superpositions of a directed and an undirected cycle.

Figure 7.1.3

The superpositions of three lines on three points.

With $m = n = 3$, the value of this product is 5. The corresponding super-positions are shown in Figure 7.1.3 with solid, dashed and dotted lines to distinguish the copies of G.

REDFIELD'S DECOMPOSITION THEOREM

The decomposition theorem for $\cup$-products is not used directly for enumeration purposes but it can be exploited in determining the cycle index of the pair group $S_p^{(2)}$ and the reduced ordered pair group $S_p^{[2]}$. Furthermore, it suggests a means of finding a formula for the cycle index sum of the groups of all graphs. The latter formula plays an important role in the enumeration of blocks as derived by Robinson [R19].

To state the theorem we first associate in a natural fashion a permutation group with each of the $N[B_1, \ldots, B_m]$ classes of matrices enumerated in (7.1.7). Suppose $[w_{ij}]$ is an $m \times n$ matrix in the kth class. Then the permutation group A_k associated with this class consists of all permutations γ such that γ is an element of each group B_i, and $[w_{ij}]$ and $[\gamma w_{ij}]$ are in the same class, i.e., are column equivalent. When Redfield's Decomposition Theorem is applied to graphs, the classes of matrices correspond to superpositions. Hence each superposition has associated with it a permutation group. Indeed, this group consists of all the permutations of the points of the superposition that preserve adjacency in each of the constituent graphs. For example, the groups of the superpositions in Figure 7.1.2 are the cyclic group C_4 and the wreath product $S_2[E_2]$.

Redfield's Decomposition Theorem   Let $B_1, \ldots, B_m$ be m permutation groups of degree n and let N be the number of classes of $m \times n$ matrices equivalent with respect to $(B_1, \ldots, B_m)$. Then the N permutation groups $H_1, \ldots, H_N$ associated with these classes satisfy

$$Z(B) \cup \cdots \cup Z(B_m) = \sum_{k=1}^{N} Z(H_k). \tag{7.2.1}$$

Thus $Z(C_4) \cup Z(D_4)$ equals the cycle index sum $Z(C_4) + Z(S_2[E_2])$ of the superpositions in Figure 7.1.2.

Proof To prove the theorem we make use of a simple group-theoretic identity, called Redfield's Lemma, which will enjoy wide applicability. Let A be a group with the permutation representation A' on object set Y. For each α in A, we denote by α' the image of α in A' for this representation. Let φ be a function from A into some polynomial ring over the rationals which is constant on the conjugate classes of A; as usual, φ is then called a *class function* for A. It is then also a class function for A'. Suppose A' has N orbits. Let $y_1, \ldots, y_N$ be elements of Y, one from each of the N orbits and for each $k = 1$ to N let

$$A_k = \{\alpha \in A \,|\, \alpha' y_k = y_k\}. \tag{7.2.2}$$

Then we have the following generalization of Burnside's Lemma:

Redfield's Lemma

$$|A|^{-1} \sum_{\alpha \in A} j_1(\alpha')\varphi(\alpha) = \sum_{k=1}^{N} |A_k|^{-1} \cdot \sum_{\alpha \in A_k} \varphi(\alpha). \tag{7.2.3}$$

The verification of (7.2.3) is similar to that of Burnside's Lemma (2.3.10) and is therefore omitted. Note that if we chose $\varphi(\alpha) = 1$ for all α, then (7.2.3) is precisely the formula in Burnside's Lemma.

For A we take all m-tuples $\alpha = (\beta_1, \ldots, \beta_m)$ with β_i in B_i for each $i = 1$ to m. Multiplication in A is defined componentwise, and α' permutes column-equivalence classes as follows. The class to which $[w_{ij}]$ belongs is sent by α' to that of $[\beta_i w_{ij}]$. Finally $\varphi(\alpha) = 0$ unless each component β_i determines the same partition (j) of n in which case $\varphi(\alpha) = \prod_{k=1}^{n} s_k^{j_k}$. The proof of (7.2.1) can now be completed by applying Redfield's Lemma (7.2.3). //

The decomposition of $\cup$-products is not necessarily unique, but there are several cases important for our purposes in which it is unique (see Redfield [R10] and especially Foulkes [F1] for a complete discussion). We are now concerned with the cycle indexes of cyclic groups. Let α be a permutation of n objects which has order r, and let $Z(\alpha)$ denote the cycle index of the cyclic group generated by α. Then we have

$$Z(\alpha) = r^{-1} \sum_{i=1}^{r} \prod_{k=1}^{n} s_{k/(k,i)}^{(k,i)j_k(\alpha)}, \tag{7.2.4}$$

and if α consists only of a single cycle of length r, Redfield's formula for $Z(\alpha)$ is easily verified:

$$Z(\alpha) = r^{-1} \sum_{d|r} \varphi(d) s_d^{r/d}, \tag{7.2.5}$$

where φ is the Euler φ-function. The following result is a consequence of the decomposition theorem and is found in Redfield [R10] in quite a different form.

Corollary If B is a permutation group of degree n and α is any permutation of n symbols which has order r, then

$$Z(B) \cup Z(\alpha) = \sum_{k|r} i_k Z(\alpha^k), \qquad (7.2.6)$$

where the i_k are uniquely determined nonnegative integers.

It follows from the Decomposition Theorem (7.2.1) that $Z(B) \cup Z(\alpha)$ is a sum of cycle indexes of groups which are subgroups of both B and the cyclic group generated by α. The coefficients i_k are unique because the cycle indexes $Z(\alpha^k)$ are independent. Furthermore, this corollary can be used to calculate the cycle index of the derived group of two permutation groups introduced next.

Let A and B have the same object set $X = \{1, \ldots, n\}$, which is also the object set of the symmetric group S_n, and let B be a subgroup of A. The *derived group* of A and B is denoted by A/B and has as its object set the right cosets of A modulo B. For each permutation α in A, there is a permutation α' in A/B such that for any right coset βB of A, the image of βB under α' is $\alpha \beta B$. That is

$$\alpha'(\beta B) = \alpha \beta B. \qquad (7.2.7)$$

Thus the permutations in A/B consist of all those permutations of the cosets which are induced by A under left multiplication. Hence A/B is a homomorphic image of A and the degree of A/B is $|A|/|B|$. The cycle index $Z(S_n/B)$ can be obtained from the next corollary.

Corollary If S_n/B is a derived group, then the permutation α' in S_n/B has i_k cycles of length k for each $k|r$, where the i_k are the coefficients of the $Z(\alpha^k)$ in the decomposition of $Z(B) \cup Z(\alpha)$.

Thus the contribution of α' to $Z(S_n/B)$ is $\prod_{k|r} s_k^{i_k}$. To illustrate, we follow Redfield and find $Z(S_5/(S_2 S_3))$ using this corollary.

From formulas (2.2.4) and (2.2.5) we have

$$Z(S_5) = (1/5!)(s_1^5 + 10s_1^3 s_2 + 20s_1^2 s_3 + 15s_1 s_2^2 + 30s_1 s_4$$

$$+ 20s_2 s_3 + 24s_5) \qquad (7.2.8)$$

$$Z(S_2)Z(S_3) = \tfrac{1}{12}(s_1^5 + 4s_1^3 s_2 + 2s_1^2 s_3 + 3s_1 s_2^2 + 2s_2 s_3). \qquad (7.2.9)$$

Now suppose α is a permutation in S_5 with cycle structure $s_2 s_3$, so that $j_2(\alpha) = j_3(\alpha) = 1$ and the order of α is 6. Since the divisors of 6 are 1, 2, 3, and 6, formula (7.2.6) implies that

$$Z(S_2)Z(S_3) \cup Z(\alpha) = i_1 Z(\alpha) + i_2 Z(\alpha^2) + i_3 Z(\alpha^3) + i_6 Z(\alpha^6). \quad (7.2.10)$$

On the other hand, since we have from (7.2.4)

$$Z(\alpha) = \tfrac{1}{6}(s_1^5 + 2s_2 s_3 + 2s_1^2 s_3 + s_1^3 s_2), \quad (7.2.11)$$

we find that

$$Z(S_2)Z(S_3) \cup Z(\alpha) = \tfrac{5}{3}s_1^5 + \tfrac{1}{3}s_2 s_3 + \tfrac{1}{3}s_1^2 s_3 + \tfrac{2}{3}s_1^3 s_2. \quad (7.2.12)$$

Combining (7.2.10) and (7.2.12) yields

$$\sum_{k \mid 6} i_k Z(\alpha^k) = \tfrac{5}{3}s_1^5 + \tfrac{1}{3}s_2 s_3 + \tfrac{1}{3}s_1^2 s_3 + \tfrac{2}{3}s_1^3 s_2. \quad (7.2.13)$$

The term $s_2 s_3$ appears in the left side of (7.2.13) only in $i_1 Z(\alpha)$. Hence the coefficient of $s_2 s_3$ in the left side of (7.2.13) is $i_1/3$. Since its coefficient in the right side of (7.2.13) is $\tfrac{1}{3}$, we have $i_1 = 1$. Subtracting $Z(\alpha)$ from both sides of (7.2.13) gives

$$i_2 Z(\alpha^2) + i_2 Z(\alpha^3) + i_6 Z(\alpha^6) = \tfrac{3}{2}s_1^5 + \tfrac{1}{2}s_1^3 s_2. \quad (7.2.14)$$

The term $s_1^2 s_3$ appears in the left side of (7.2.14) only in $i_2 Z(\alpha^2)$ and since it does not appear at all on the right side, we must have $i_2 = 0$. Now $s_1^3 s_2$ has coefficient $i_3/2$ on the left side of (7.2.14) and $\tfrac{1}{2}$ on the right, so $i_3 = 1$. Subtracting $Z(\alpha^3)$ from both sides of (7.2.14) leaves

$$i_6 Z(\alpha^6) = s_1^5,$$

and hence $i_6 = 1$. Thus the cycle structure of α' in $S_5/(S_2 S_3)$ is given by $s_1 s_3 s_6$. Therefore the 20 permutations with structure $s_2 s_3$ in S_5 contribute $(20/5!)s_1 s_3 s_6$ to $Z(S_5/(S_2 S_3))$. Next we observe that α^2 has structure $s_1^2 s_3$ and since $(\alpha^2)' = (\alpha')^2$, the 20 permutations with structure $s_1^2 s_3$ contribute $(20/5!)s_1 s_3^3$ to the cycle index of the derived group. Similarly, since α^3 has structure $s_1^3 s_2$, we have the term $(10/5!)s_1^4 s_2^3$. We can continue in this manner to determine the contributions to the derived group of other permutations in S_5 by selecting other elements which generate maximal cyclic subgroups. On completing the process we have

$$Z(S_5/(S_2 S_3)) = (1/5!)(s_1^{10} + 10s_1^4 s_2^3 + 15s_1^2 s_2^4 + 20s_1 s_3^3$$
$$+ 20s_1 s_3 s_6 + 30s_2 s_4^2 + 24s_5^2). \quad (7.2.15)$$

Redfield (see [R10, pp. 451–453]) used this procedure in order to calculate $Z(S_n/(S_2 S_{n-1}))$ for $n \le 7$.

Next we show that Redfield's Lemma (7.2.3) may be used to obtain Robinson's formula [R19] for the cycle index sum of the groups of all graphs. We will see in the next chapter that formula (7.2.18), to be derived now, is of crucial importance in the counting of blocks. As usual S_p is the symmetric group acting on $X = \{1, \ldots, p\}$ and so the pair group $S_p^{(2)}$ has object set $X^{(2)}$, the collection of 2-subsets of X. We denote by $S_p^{(1,2)}$ the group induced by S_p which acts on the union $X \cup X^{(2)}$. For $p > 2$, both of these groups are abstractly isomorphic to S_p. In the formula for the cycle index of $S_p^{(1,2)}$ we use two sets of variables, s_k and t_k, to distinguish between cycles of 1-subsets and of 2-subsets respectively (s for single and t for two). This distinction is indicated by writing $Z(S_p^{(1,2)}; s_k, t_k)$ for the cycle index of $S_p^{(1,2)}$. Therefore if α in S_p induces α' in $S_p^{(2)}$, then the contribution of α to $Z(S_p^{(1,2)})$ can be written

$$\left(\prod_{k=1}^{p} s_k^{j_k(\alpha)} \right) \left(\prod_{k=1}^{r} t_k^{j_k(\alpha')} \right)$$

where $r = \binom{p}{2}$.

Let E_2 be the identity permutation group acting on $Y = \{0, 1\}$. The class function φ for S_p is defined by

$$\varphi(\alpha) = \prod s_k^{j_k(\alpha)}. \tag{7.2.16}$$

We apply Redfield's Lemma (7.2.3) to the representation $E_2^{S_p^{(2)}}$ of S_p. Then the right side of formula (7.2.2) is $\sum Z(G)$, where the sum is over all (nonisomorphic) graphs with p points.

For each permutation of the form $(\alpha'; (0)(1))$ in $E_2^{S_p^{(2)}}$, we also have

$$j_1(\alpha'; (0)(1)) = \sum_{k=1}^{r} 2^{j_k(\alpha')}. \tag{7.2.17}$$

Therefore the left side of (7.2.3) is

$$(1/p!) \sum_{\alpha \in S_p} \prod_{k=1}^{r} 2^{j_k(\alpha')} \prod_{k=1}^{p} s_k^{j_k(\alpha)}.$$

But this last expression is simply $Z(S_p^{(1,2)}); s_k, 2)$, and we have the result of Robinson [R19].

Theorem The cycle index sum for all graphs with p points is

$$\sum Z(G) = Z(S_p^{(1,2)}; s_k, 2). \tag{7.2.18}$$

A variation of (7.2.18) is obtained when each t_k in $Z(S_p^{(1,2)})$ is replaced by $1 + t_k$. We then have the cycle index sum of the point-line groups $\Gamma_{0,1}(G)$ for all graphs with p points:

$$\sum Z(\Gamma_{0,1}(G)) = Z(S_p^{(1,2)}); s_k, 1 + t_k. \tag{7.2.19}$$

7.3 GRAPHS AND DIGRAPHS

We shall indicate in this section how graphs and digraphs may be regarded as "superpositions" and therefore may be enumerated by Redfield's Enumeration Theorem. First we observe that (7.1.7) can be used to count certain 1–1 functions. Recall from Section 6.6 that B^{A*} denotes the restriction of the power group B^A to 1–1 functions.

Corollary If A and B are permutation groups of degree n, then the number of orbits determined by the restricted power group B^{A*} is $Z(A) \cap Z(B)$.

The proof follows from (7.1.7) with $m = 2$, because each $2 \times n$ column-equivalence class of matrices is a 1–1 function from one object set to the other. Such column-equivalence classes are equivalent with respect to A and B if and only if the corresponding 1–1 functions are in the same orbit of B^{A*}.

$$//$$

We now seek the number $g_{p,q}$ of graphs with p points and q lines. To this end let the pair group $S_p^{(2)}$ have object set $X^{(2)}$ with $X = \{1, \ldots, p\}$ as usual. Also let $Y = Y_1 \cup Y_2$ be a set of $r = \binom{p}{2}$ integers with $Y_1 = \{1, \ldots, q\}$ and $Y_2 = \{q + 1, \ldots, r\}$. If group S_q has object set Y_1 and group S_{r-q} has object set Y_2, we can form the product $S_q S_{r-q}$, which has cycle index $Z(S_q)Z(S_{r-q})$. Now with $A = S_p^{(2)}$ and $B = S_q S_{r-q}$, it can be seen that the orbits of the restricted power group constitute the isomorphism classes of graphs with p points and q lines. Each 1–1 function $f : X^{(2)} \to Y$ represents a graph with point set X, in which i and j are adjacent if $f\{i, j\}$ is an element of Y_1. Clearly these functions are in the same orbit of B^{A*} if and only if they represent isomorphic graphs. Hence, as Redfield observed, graphs may be enumerated using $\cap$-products.

Theorem The number of (p, q) graphs is

$$g_{p,q} = Z(S_p^{(2)}) \cap Z(S_q)Z(S_{r-q}), \tag{7.3.1}$$

where $r = \binom{p}{2}$.

Redfield was able to compute the numbers of graphs with $p \leq 7$ having no isolated points using a formula similar to (7.3.1). In the terminology of Whitehead and Russell [WR1] he referred to these graphs as "symmetrical aliorelative dyadic relations on fields of p elements." He had no explicit formula, such as (4.1.9) for $Z(S_p^{(2)})$, but since the pair group can be realized as a derived group, he was able to employ the corollaries of his Decomposition Theorem to obtain the appropriate cycle index formula for $Z(S_p^{(2)})$. Specifically, we find the following identity for the pair group as a derived group:

$$S_p^{(2)} = S_p/(S_2 S_{p-2}). \tag{7.3.2}$$

Although we omit its proof, note that the degree of each side is $\binom{p}{2}$ and see formula (7.2.15) for $Z(S_5/(S_2 S_3))$. Thus the Decomposition Theorem was a basic part of Redfield's enumeration techniques. We note briefly the corresponding formulas for directed graphs. Since

$$S_p^{[2]} = S_p/(E_2 S_{p-2}), \tag{7.3.3}$$

the cycle index $Z(S_p^{[2]})$ of the reduced ordered pair group can be found using the two corollaries of the Decomposition Theorem. Then the number d_{pq} of digraphs with p points and q lines is

$$d_{p,q} = Z(S_p^{[2]}) \cap Z(S_q) Z(S_{r-q}), \tag{7.3.4}$$

where $r = p(p-1)$.

7.4   A GENERALIZATION OF REDFIELD'S ENUMERATION THEOREM

In this section we shall construct a more general permutation group than that required in the proof of Redfield's Enumeration Theorem, and give an explicit formula for the number of orbits in it. This result enables us to enumerate superposed graphs composed of interchangeable copies of the same graph (see Palmer and Robinson [PR2] and [PR3]), and it also provides a new approach to the enumeration of multigraphs [P4].

Let A and B be permutation groups with object sets $X = \{1, \ldots, m\}$ and $Y = \{1, \ldots, n\}$ respectively, and let W be the collection of $m \times n$ matrices in which the elements of each row are the n objects in Y. The *matrix group* $[A; B]$ of A and B acts on the column-equivalence classes of W as follows. For each permutation α in A and each sequence $\beta_1, \beta_2, \ldots, \beta_m$ of m permutations with β_i in B, there is a permutation, denoted $[\alpha; \beta_1, \beta_2, \ldots, \beta_m]$ in $[A; B]$ such that the column-equivalence class to which the matrix

$[w_{ij}]$ belongs is sent by $[\alpha; \beta_1, \beta_2, \ldots, \beta_m]$ to the class to which $[\beta_i w_{\alpha i, j}]$ belongs. That is, α first determines a permutation of the rows and then each β_i permutes the entries in the i'th row. For example, let $A = S_2$ and $B = S_2[E_2]$ so that $[S_2; S_2[E_2]]$ permutes 2×4 matrices. With $\alpha = (12)$, $\beta_1 = (1)(2)(3)(4)$ and $\beta_2 = (12)(34)$ it follows that $[\alpha; \beta_1, \beta_2]$ sends the class of the matrix

$$\begin{pmatrix} 1 & 2 & 3 & 4 \\ 3 & 1 & 4 & 2 \end{pmatrix} \quad \text{to that of} \quad \begin{pmatrix} 3 & 1 & 4 & 2 \\ 2 & 1 & 4 & 3 \end{pmatrix}.$$

From the definition it follows that with $A = E_m$, the identity group, the number $N[E_m; B]$ of orbits of the matrix group is the cap-product of m copies of $Z(B)$. Therefore if B is the group of a graph G of order n, $N[E_m; \Gamma(G)]$ is the number of superpositions of m copies of G. On the other hand, $N[S_m; \Gamma(G)]$ is the number of superpositions in which the copies of G are interchangeable. To illustrate, we consider the path P_4 of order 4. All eight superposed graphs composed of two copies of P_4 are shown in Figure 7.4.1. Solid and dashed lines are used to distinguish the copies. Interchanging the solid and dashed lines permutes the last two graphs in the figure and leaves each of the first six fixed. Thus the figure confirms that $N[E_2; \Gamma(P_4)] = 8$ and $N[S_2; \Gamma(P_4)] = 7$; this latter number is calculated in (7.4.13) below.

In order to state the theorem which gives the formula for the number $N[A; B]$ of orbits of the matrix group $[A; B]$, we require several definitions. As usual R is the ring of polynomials in the variables $s_1, s_2, \ldots, s_n$. For each positive integer r, we define a function $J_r: R \to R$. It is convenient first to define a sequence of functions $d_1, d_2, \ldots$ which depend on the integers r and $k \geq 1$. For each i, let

$$d_i = \begin{cases} s_{ki}/k & \text{if } i | r \text{ and } (r/i, k) = 1 \\ 0 & \text{otherwise.} \end{cases} \tag{7.4.1}$$

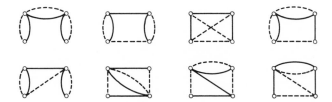

Figure 7.4.1

The superpositions of two paths of order 4.

Using S_j as usual to denote the symmetric group on j objects we define $J_r(s_k^j)$ by

$$J_r(s_k^j) = j! \, k^j Z(S_j; d_1, d_2, \ldots, d_j). \tag{7.4.2}$$

Observe that for any prime p, the previous formula (7.4.2) for $J_p(s_k^j)$ can be written

$$J_p(s_k^j) = \begin{cases} 0 & \text{if } p|k \text{ but } p \nmid j \\ (j! \, k^{j(p-1)/p} s_{pk}^{j/p})/((j/p)! \, p^{j/p}), & \text{if } p|k \text{ and } p|j \\ \sum_{t=0}^{[j/p]} (j! \, k^{(p-1)t} s_{kp}^t s_k^{j-tp})/((j-tp)! \, t! \, p^t) & \text{if } p \nmid k. \end{cases} \tag{7.4.3}$$

For monomials $s_1^{j_1} s_2^{j_2} \cdots s_n^{j_n}$, we then define J_r by

$$J_r(s_1^{j_1} s_2^{j_2} \cdots s_n^{j_n}) = \prod_{k=1}^{n} J_r(s_k^{j_k}). \tag{7.4.4}$$

Now J_r is extended linearly to R. In particular

$$J_r(Z(B)) = |B|^{-1} \sum_{\beta \in B} J_r\left(\prod_{k=1}^{n} s_k^{j_k(\beta)} \right) \tag{7.4.5}$$

and

$$J_1(Z(B)) = Z(B). \tag{7.4.6}$$

Next we define a product for the collection $\{J_r\}$ of functions. We set

$$J_1^{i_1} J_2^{i_2} \cdots J_m^{i_m}(Z(B)) = J_1^{i_m}(Z(B)) \cup \cdots \cup J_m^{i_m}(Z(B)) \tag{7.4.7}$$

where it is understood that before evaluating the right side of (7.4.7), each $J_r^{i_r}(Z(B))$ is replaced by the cup-product of length i_r,

$$J_r(Z(B)) \cup \cdots \cup J_r(Z(B)).$$

Theorem The number of orbits $N[A; B]$ determined by the matrix group $[A; B]$ is

$$N[A; B] = [Z(A; J_1, J_2, \ldots, J_m) Z(B)]_{s_k = 1}. \tag{7.4.8}$$

We now apply the theorem to obtain the number $N[S_2; \Gamma(P_4)]$ of superposed graphs composed of two interchangeable copies of P_4.

The cycle indexes of $\Gamma(P_4)$ and S_2 are

$$Z(\Gamma(P_4)) = \tfrac{1}{2}(s_1^4 + s_2^2) \tag{7.4.9}$$

and

$$Z(S_2; J_1, J_2) = \tfrac{1}{2}(J_1^2 + J_2). \tag{7.4.10}$$

From (7.4.6) we have $J_1(Z(\Gamma(P_4))) = Z(\Gamma(P_4))$. And from (7.4.3) we obtain

$$J_2(Z(\Gamma(P_4))) = \tfrac{1}{2}(s_1^4 + 6s_1^2 s_2 + 3s_2^2 + 2s_4). \tag{7.4.11}$$

It is easily seen that

$$Z(S_2; J_1, J_2)(\tfrac{1}{2}(s_1^4 + s_2^2)) = \tfrac{1}{2}\{\tfrac{1}{4}(24s_1^4 + 8s_2^2)$$
$$+ \tfrac{1}{2}(s_1^4 + 6s_1^2 s_2 + 3s_2^2 + 2s_4)\}. \tag{7.4.12}$$

Evaluating this expression with $s_k = 1$ for all k gives

$$N[S_2; \Gamma(P_4)] = 7, \tag{7.4.13}$$

which is verified in Figure 7.4.1.

TABLE 7.4.1

SUPERPOSITIONS OF CYCLES

n	$N[E_2; D_n]$	$N[S_2; D_n]$	$N[E_3; D_n]$	$N[S_3; D_n]$
3	1	1	1	1
4	2	2	5	3
5	4	4	24	9
6	12	10	391	89

The number of superpositions of two and three cycles of order ≤ 6 are given in Table 7.4.1. Note that there are two superpositions of two cycles of order 6 which are equivalent when the constituent cycles are interchangeable.

The number m_{pq} of multigraphs with p points and q lines can also be calculated using the matrix group. Let G be the disjoint union $K_2 \cup \bar{K}_{p-2}$. Then the superpositions of q interchangeable copies of G correspond precisely to these multigraphs. Since $\Gamma(G) = S_2 S_{p-2}$ we can express the formula from [P4] as follows.

Theorem The number $m_{p,q}$ of multigraphs with p points and q lines is given by

$$m_{p,q} = [Z(S_q; J_1, \ldots, J_q)(Z(S_2)Z(S_{p-2}))]_{s_k = 1}. \tag{7.4.14}$$

Some of the details in finding $m_{5,4}$ are now sketched. First of all we have the product $Z(S_2)Z(S_3)$ in (7.2.9) and $Z(S_4; J_1, J_2, J_3, J_4)$ is obtained from

(4.1.10). From equations (7.4.1) through (7.4.5) we have the following calculations:

$$J_2(Z(S_2)Z(S_3)) = \tfrac{1}{12}(s_1^5 + 10s_1^3s_2 + 15s_1s_2^2 + 2s_1^2s_3 + 2s_2s_3 + 6s_1s_4)$$

$$J_3(Z(S_2)Z(S_3)) = \tfrac{1}{12}(s_1^5 + 20s_1^2s_3 + 4s_1^3s_2 + 8s_2s_3 + 3s_1s_2^2) \tag{7.4.15}$$

$$J_4(Z(S_2)Z(S_3)) = \tfrac{1}{12}(s_1^5 + 10s_1^3s_2 + 15s_1s_2^2 + 30s_1s_4 + 2s_1^2s_3 + 2s_2s_3).$$

From the definition (7.4.7) of products, we find

$$[J_1^4(Z(S_2)Z(S_3))]_{s_k=1} = 107$$

$$[6J_1^2J_2(Z(S_2)Z(S_3))]_{s_k=1} = 162$$

$$[8J_1J_3(Z(S_2)Z(S_3))]_{s_k=1} = 40 \tag{7.4.16}$$

$$[3J_2^2(Z(S_2)Z(S_3))]_{s_k=1} = 69$$

$$[6J_4(Z(S_2)Z(S_3))]_{s_k=1} = 30.$$

Therefore $m_{5,4} = \tfrac{1}{24}(107 + 162 + 40 + 69 + 30) = 17$. The great advantages of this method are that only the cycle index formulas of the symmetric groups are required and $m_{p,q}$ is computed directly.

7.5 GENERAL GRAPHS

A *general graph* is permitted to have both multiple lines and multiple loops. By definition each loop at a point contributes 2 to the degree of that point. In this section we shall show how Read [R1] was able to enumerate general graphs with specified degree sequence by finding a 1–1 correspondence between these graphs and certain superpositions.

First, we require a formula for the number of bicolored multigraphs, that is multigraphs whose points are partitioned into two sets, one consisting of red points, the other of blue points, and every line of the graph joins a red point with a blue one. The two bicolored graphs with two red points of degree 3 and three blue points of degree 2 are shown in Figure 7.5.1. The following theorem provides a formula for the number of bicolored graphs with specified degrees.

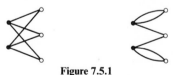

Figure 7.5.1

Two bicolored graphs.

Theorem   The number of bicolored graphs which have for each positive integer k exactly r_k red points of degree k and b_k blue points of degree k is

$$Z\left(\prod_k S_{r_k}[S_k]\right) \cap Z\left(\prod_k S_{b_k}[S_k]\right). \tag{7.5.1}$$

For example, it is easily shown that $Z(S_3[S_2]) \cap Z(S_2[S_3]) = 2$ which agrees with Figure 7.5.1.

Proof   Let G_1 be a graph that has for each k, r_k components which are complete graphs on k points. Similarly, G_2 has for each k, b_k components which are complete graphs on k points. Note that the number of lines in the bicolored graphs to be counted is $q = \sum_k kr_k = \sum_k kb_k$. Therefore G_1 and G_2 have the same number of points, namely q. If the components of G_1 are interpreted as red points and the components of G_2 as blue points, then each superposition of G_1 and G_2 is seen to correspond to a bichromatic graph. This correspondence is clearly 1–1 and therefore the number of bicolored graphs is $Z(G_1) \cap Z(G_2)$. Furthermore $\Gamma(G_1)$ and $\Gamma(G_2)$ can be expressed as a direct product of wreath products, as in (7.5.1).   **//**

Next we seek the number of general graphs that have for each k, d_k points of degree k. The number of lines in these graphs is, therefore, $q = \frac{1}{2}\sum kd_k$. Suppose the points of such a graph are colored red and then a blue point is inserted on each line. A bicolored graph is then obtained in which all q blue points have degree 2. This correspondence indicated between general graphs and bicolored graphs is, of course, one-to-one and thus we have the next result.

Corollary   The number of general graphs which have for each k, d_k points of degree k is

$$Z\left(\prod_k S_{d_k}[S_k]\right) \cap Z(S_q[S_2]). \tag{7.5.2}$$

For example, the number of general graphs with two points of degree 3 is $Z(S_2[S_3]) \cap Z(S_3[S_2]) = 2$. The two general graphs enumerated can be obtained by suppressing the points of degree 2 in Figure 7.5.1.

Regular general graphs and cubic general graphs are obtained at once as special cases.

Corollary   The number of general graphs on n points which are regular of degree k is

$$Z(S_n[S_k]) \cap Z(S_{nk/2}[S_2]). \tag{7.5.3}$$

Corollary   The number of general cubic graphs on $2n$ points is

$$Z(S_{2n}[S_3]) \cap Z(S_{3n}[S_2]). \tag{7.5.4}$$

In the formulas just above we have been concerned with enumerating unlabeled graphs. The superposition theory has been exploited with great success by Read [R1] to enumerate various types of labeled graphs. For example, it follows quickly from (7.5.4) that the number of general cubic graphs with $2n$ labeled points is

$$Z(E_{2n}[S_3]) \cap Z(S_{3n}[S_2]). \tag{7.5.5}$$

Our next goal, then, is to find formulas for labeled cubic graphs with no loops and no multiple lines. The result will be expressions remarkably similar to (7.5.4). We shall provide, first, general results from which the cubic case follows immediately.

As in (2.6.1), we denote the difference $Z(A_k) - Z(S_k)$ by $Z(A_k - S_k)$. We shall also make use of the notation provided in Chapter 4 for the cycle index of the wreath product. In particular, if P_1 and P_2 are polynomials in the variables $s_1, s_2, \ldots$, then $P_1[P_2]$ denotes that polynomial obtained by replacing each variable s_k in P_1 by the polynomial which results when one multiplies all the subscripts of the variables in P_2 by k. Thus expressions such as $Z(S_q)[Z(A_2 - S_2)]$ and $Z(S_{3n})[Z(A_2 - S_2)]$ are defined.

The next three theorems are obtained by applying PET with suitable group and figure counting series, and by expressing the appropriate coefficient as a $\cap$-product; see Read [R1] for the details. These three results and their respective corollaries count the three kinds of labeled graphs with given partition and labeled cubic graphs: (1) multigraphs, (2) general graphs with no multiple lines, and (3) graphs.

Theorem   The number of labeled multigraphs with d_k points of degree k and hence $q = \frac{1}{2}\sum k d_k$ lines is

$$Z\left(\prod_k E_{d_k}[S_k]\right) \cap Z(S_q)[Z(A_2 - S_2)]. \tag{7.5.6}$$

We now apply this theorem to cubic graphs with $2n$ points to get the next formula.

Corollary   The number of labeled, cubic multigraphs with $2n$ points is

$$Z(E_{2n}[S_3]) \cap Z(S_{3n})[Z(A_2 - S_2)]. \tag{7.5.7}$$

On excluding multiple lines but not loops we have the next result.

Theorem The number of labeled, general graphs without multiple lines and with d_k points of degree k and q lines is

$$Z\left(\prod_k E_{d_k}[S_k]\right) \cap Z(A_q - S_q)[Z(S_2)]. \tag{7.5.8}$$

When this is applied to cubic graphs, we arrive at the next formula.

Corollary The number of labeled, general, cubic graphs with $2n$ points and no multiple lines is

$$Z(E_{2n}[S_3]) \cap Z(A_{3n} - S_{3n})[Z(S_2)]. \tag{7.5.9}$$

Finally, both loops and multiple lines are excluded.

Theorem The number of labeled graphs with d_k points of degree k and q lines is

$$Z\left(\prod_k E_{d_k}[S_k]\right) \cap Z(A_q - S_q)[Z(A_2 - S_2)]. \tag{7.5.10}$$

Corollary The number of labeled, cubic graphs with $2n$ points and hence $3n$ lines is

$$Z(E_{2n}[S_3]) \cap Z(A_{3n} - S_{3n})[Z(A_2 - S_2)]. \tag{7.5.11}$$

To evaluate the expressions (7.5.7), (7.5.9), and (7.5.11) requires a considerable amount of computation, but Read was able to obtain some general results for these in [R1]. For example, in the case of (7.5.11) he found:

Corollary The number of labeled, cubic graphs with $2n$ points is

$$\frac{(2n)!}{6^n} \sum_{j,k} \frac{(-1)^j(6k - 2j)!\,6^j}{(3k - j)!(2k - j)!j!(n - k)!} 48^k \sum_i \frac{(-1)^j j!}{(j - 2i)!i!} \tag{7.5.12}$$

and the number of these is asymptotic to

$$\frac{(6n)!\,e^{-2}}{288^n(3n)!}. \tag{7.5.13}$$

An extensive discussion of the problem of evaluating cap-products and of methods using S-functions is given by Read [R8]. Further asymptotic results are developed in Chapter 9.

EXERCISES

7.1 How many superpositions can be obtained from three directed cycles of order 6?

7.2 Find two permutation groups A and B such that the decomposition of $Z(A) \cup Z(B)$ as a sum of cycle indexes is not unique. (Redfield [R10])

7.3 Express $Z(A; J_1, \ldots, J_m)Z(B)$ as a sum of $N[A; B]$ cycle indexes.
(Palmer and Robinson [PR3])

7.4 How many orbits of 1–1 functions are determined by the restricted power groups $E_n^{E_m*}$, $E_n^{S_m*}$, and $S_n^{E_m*}$?

7.5 The pair group $S_p^{(2)}$ and the derived group $S_p/(S_2 S_{p-2})$ are identical.

7.6 Calculate $Z(S_6/(S_2 S_4))$ using (7.2.6).

7.7 How many superposed graphs can be formed from three interchangeable copies of a directed cycle of order 6?

7.8 Calculate the number of (a) general cubic graphs (b) labeled cubic graphs of order 4, 6, and 8.

7.9 Draw the 12 superpositions of two cycles of order 6 and the 10 with interchangeable cycles.

7.10 The coefficient of x^m in $Z(S_n/B, 1/(1 - x))$ is $N[S_m; B]$. Hence the generating function for superpositions with interchangeable colors of any graph G of order p is $Z(S_p/\Gamma(G), 1/(1 - x))$.

Chapter 8 | BLOCKS

The theoretical physicist, G. E. Uhlenbeck, in the Gibbs Lecture entitled "Unsolved problems in statistical mechanics," given at a meeting of the American Mathematical Society in 1950, cited the enumeration of blocks as one of these problems. Subsequently Riddell [R14], and Ford and Uhlenbeck [FU1] counted labeled blocks (Section 1.3), but it was Robinson [R19] who succeeded in solving the unlabeled problem.

Since we have already enumerated connected graphs in Chapter 4, it is sufficient for counting blocks to find the number of connected graphs that *do* have at least one cutpoint. Ordinary generating functions do not carry enough information about the structure of graphs, and so the appropriate basic technique involves sums of cycle indexes of groups of graphs (Section 7.2.). This approach was implicit in Redfield's Decomposition Theorem [R10] and was fully developed in Robinson's solution to this problem. Robinson's method can also be used to enumerate graphs with given blocks, connected graphs with no endpoints, acyclic digraphs, and other kinds of graphs. The material in this chapter is essentially due to Robinson.

177

8.1   A GENERALIZATION OF REDFIELD'S LEMMA

To prove the Composition Theorem of Section 8.3, we need a slight generalization of Redfield's Lemma (7.2.3). Let A be a permutation group with object set X and let R be any commutative ring which contains the rationals. Let φ be any function from the cartesian product $A \times X$ into R which satisfies the following condition for α and β in A and x in X:

$$\text{If } \alpha x = \beta y = x, \quad \text{then} \quad \varphi(\alpha, x) = \varphi(\beta \alpha \beta^{-1}, y). \tag{8.1.1}$$

The orbit of x determined by A is denoted by $\tilde{x}$ and $A(x) = \{\alpha \in A \,|\, \alpha x = x\}$. Then the new lemma takes the following form.

Lemma   For any function φ satisfying (8.1.1), the following identity holds:

$$|A|^{-1} \sum_{\alpha \in A} \sum_{x = \alpha x} \varphi(\alpha, x) = \sum_{\tilde{x}} |A(x)|^{-1} \sum_{\alpha \in A(x)} \varphi(\alpha, x). \tag{8.1.2}$$

The proof is made as in (7.2.3) by interchanging the sums on the left side of (8.1.2). Condition (8.1.1) allows us then to sum over the orbits of A in X, instead of all the elements of X.   **//**

Note that (8.1.2) holds in the case in which X is infinite but countable provided that all of the sums involved are defined. If ψ is a class function for A, then $\varphi(\alpha, x) = \psi(\alpha)$ satisfies (8.1.1) and with this definition (8.1.2) becomes Redfield's Lemma.

8.2   THE COMPOSITION GROUP

To state the Composition Theorem of Section 8.3, it is necessary to generalize the definition of the composition or wreath product of permutation groups. We also provide a formula for the cycle index of this group, thereby generalizing Pólya's formula (4.3.15).

As observed in [H1, p. 166] any graph G can be written as $G = n_1 G_1 \cup n_2 G_2 \cup \cdots \cup n_r G_r$, where n_i is the number of components of G isomorphic to G_i. Then the group of G is expressed as a product of composition groups:

$$\Gamma(G) = \prod_{k=1}^{r} S_{n_k}[\Gamma(G_k)]. \tag{8.2.1}$$

For example, the graph $2K_{1,2} \cup 3K_2$ of Figure 8.2.1 has as its group

$$\Gamma(2K_{1,2} \cup 3K_3) = S_2[S_1 S_2](S_3[S_2]). \tag{8.2.2}$$

Figure 8.2.1

The graph $2K_{1,2} \cup 3K_2$.

Let $G_1 = K_{1,2}$ and $G_2 = K_2$. Then each function f from $X = \{1, 2, 3, 4, 5\}$ into $Y = \{1, 2\}$ corresponds in a natural way to a graph whose components are $K_{1,2}$ and K_2. For example, any function which sends two elements of X to 1 and the other three to 2 corresponds to $2G_1 \cup 3G_2$. Suppose $f(1) = f(2) = 1$ and $f(3) = f(4) = f(5) = 2$. Then the permutations α of X such that

$$f(x) = f(\alpha x) \qquad (8.2.3)$$

for all x in X, constitute a group identical to the product $S_2 S_3$. This group is denoted by $S_5(f)$ because it consists of the permutations in S_5 that fix f, and it determines how the components of $2G_1 \cup 3G_2$ may be permuted among themselves. With this example in mind, we now generalize the composition group.

Let A be a permutation group with object set $X = \{1, 2, \ldots, n\}$ and let f be a function from X onto $Y = \{1, 2, \ldots, m\}$. Then $A(f)$ is the subgroup of A defined by

$$A(f) = \{\alpha \in A \mid \text{for each } x \text{ in } X, \quad f(x) = f(\alpha x)\}. \qquad (8.2.4)$$

Therefore if we set $X_i = f^{-1}(i)$ for each i in Y, then $\alpha(X_i) = X_i$ for each i in Y and α in $A(f)$. For each $i = 1$ to m let B_i be a permutation group with object set Y_i. Then the *generalized composition group*, denoted $A(f)[B_1, \ldots, B_m]$, has its object set the union $\bigcup_{i=1}^{m} X_i \times Y_i$. For each α in $A(f)$ and each sequence $\tau_1, \ldots, \tau_m$ of functions such that $\tau_i : X_i \to B_i$ for $i = 1$ to m so that $t_i(x)$ is a permutation in B_i, there is a permutation in $A(f)[B_1, \ldots, B_m]$, denoted $(\alpha; \tau_1, \ldots, \tau_m)$, which permutes the ordered pairs (x, y) in the object set according to the rule:

$$(\alpha; \tau_1, \ldots, \tau_m)(x, y) = (\alpha x, \tau_i(x)y) \qquad (8.2.5)$$

whenever (x, y) is in $X_i \times Y_i$. One can then easily verify the fact that the permutations of this form are closed under multiplication so that the collection does constitute a group. Intuitively, $A(f)[B_1, \ldots, B_m]$ permutes copies of the sets Y_i for $i = 1$ to m, with one copy for each element of X_i. The copies of Y_i are permuted among themselves by the permutations in $A(f)$, while the elements of each copy are permuted independently by B_i.

Since for each i we have $\alpha(X_i) = X_i$, for each k we can define $j_k(\alpha, X_i)$ to be the number of cycles of length k in α restricted to X_i. Then the generalization of (4.3.15) may be stated as follows.

Theorem   The cycle index of the generalized composition group is given by

$$Z(A(f)[B_1, \ldots, B_m]) = |A(f)|^{-1} \sum_{\alpha \in A(f)} \prod_{i=1}^{m} \prod_{k} Z(B_i; s_k, s_{2k}, s_{3k}, \ldots)^{j_k(\alpha, X_i)}$$

$$(8.2.6)$$

Note that when $m = 1$, then $Z(A(f)[B_1]) = Z(A(f))[Z(B_1)]$ and formula (8.2.6) reduces to (4.3.15). The proof of (4.3.15) can be adapted to a proof of (8.2.6).

To illustrate the theorem, we return to the example above where $A = S_5$, $f(1) = f(2) = 1$, and $f(3) = f(4) = f(5) = 2$. Then $X_1 = \{1, 2\}$ and $X_2 = \{3, 4, 5\}$. We take $B_1 = \Gamma(K_{1,2}) = S_1 S_2$ and $B_2 = \Gamma(K_2) = S_2$ with object sets $Y_1 = \{1, 2, 3\}$ and $Y_2 = \{1, 2\}$ respectively. Then the composition group $S_5(f)[S_1 S_2, S_2]$ has $(X_1 \times Y_1) \cup (X_2 \times Y_2)$ as its object set, and the ordered pairs in this set correspond to the points of $2K_{1,2} \cup 3K_2$ as indicated in Figure 8.2.2.

Each function $\tau_1 : X_1 \to \Gamma(K_{1,2})$ associates with each copy of $K_{1,2}$ a permutation in $\Gamma(K_{1,2})$ and $\tau_2 : X_2 \to \Gamma(K_2)$ does the same for K_2. Suppose $(\alpha; \tau_1, \tau_2)$ is an element of $S_5(f)[S_1 S_2, S_2]$. From (8.2.5) it follows that $\tau_1(1)$ and $\tau_1(2)$ permute the points of the two copies of $K_{1,2}$ while $\tau_2(3)$, $\tau_2(4)$ and $\tau_2(5)$ permute the points of the three copies of K_2. Then α permutes the components of $2K_{1,2} \cup 3K_2$ among themselves. Thus it is seen that

$$\Gamma(2K_{1,2} \cup 3K_2) = S_5(f)[S_1 S_2, S_2] \qquad (8.2.7)$$

Finally, from the formula (8.2.6) for the cycle index it can be shown that

$$Z(S_5(f)[S_1 S_2, S_2]) = (1/2! \, 3!)\{Z(S_1 S_2)^2 + Z(S_1 S_2; s_2, s_4)\}$$

$$\{Z(S_2)^3 + 3Z(S_2)Z(S_2; s_2, s_4) + 2Z(S_2; s_3, s_6)\}.$$

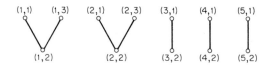

Figure 8.2.2

The graph $2K_{1,2} \cup 3K_2$ with point sets $(X_1 \times Y_1) \cup (X_2 \times Y_2)$.

This equation is easily verified by applying the cycle index formulas (4.3.14) and (2.2.14) to (8.2.2).

8.3 THE COMPOSITION THEOREM

For any set $\mathscr{H}$ of graphs, we shall denote by $Z(\mathscr{H})$ the cycle index sum of the groups of the graphs in $\mathscr{H}$; compare (7.2.18), the cycle index sum of the groups of all graphs of order p. Under special circumstances each of these groups may be a composition group formed from a subgroup of the group A and a sequence of groups $B_1, B_2, B_3, \ldots$. We shall show here how to obtain $Z(\mathscr{H})$, given $Z(A)$ and $Z(B_i)$ for each $i = 1, 2, 3, \ldots$.

In particular, let A be a permutation group with object set $X = \{1, \ldots, n\}$. Let Y be any nonempty subset of the positive integers, and let B_i be a permutation group with object set Y_i for each i in Y. In practice, each element i of Y corresponds to a graph with group B_i and point set Y_i. We require throughout this discussion that only finitely many Y_i have the same cardinality, so that the sum $\sum Z(B_i)$ over all i in Y is defined.

For each function $f : X \to Y$, we define the subgroup $A(f)$ of A as above. Next we define an equivalence relation for these functions from X to Y. We say that f and g are *equivalent with respect to A*, and we write $f \sim g$ if for some α in A

$$f(x) = g(\alpha x) \tag{8.3.1}$$

for all x in X; that is, $f \sim g$ means that f and g are in the same orbit of the power group E^A. We denote the set of equivalence classes (orbits) by $\mathscr{F}$. Now each function f has associated with it a generalized composition group. If $f(X) = \{i_1, \ldots, i_m\}$, then this group formed by $A(f)$ and $B_{i_1}, \ldots, B_{i_m}$ can be denoted by $A(f)[B_{i_1}, \ldots, B_{i_m}]$. If f and g and equivalent, then $A(f)$ and $A(g)$ are identical permutation groups. In fact, they are conjugate because if $f(x) = g(\gamma x)$ for some γ in A and all x in X, then $\gamma A(f) \gamma^{-1} = A(g)$. Furthermore, the generalized composition groups $A(f)[B_{i_1}, \ldots, B_{i_m}]$ and $A(g)[B_{i_1}, \ldots, B_{i_m}]$ are also identical. Hence they have the same cycle index, and we can define the cycle index of any equivalence class of functions F in $\mathscr{F}$ to be the cycle index of the composition group determined by any function f in F. If we let $Z(\mathscr{F}) = \sum_{F \in \mathscr{F}} Z(F)$, then the main result, expressed in the next theorem, relates $Z(\mathscr{F})$ to $Z(A)$ and the $Z(B_i)$. Recall that, by $Z(A)[\sum Z(B_i)]$ we mean the power series obtained when each variable s_k in $Z(A)$ is replaced by

$$\sum Z(B_i; s_k, s_{2k}, s_{3k}, \ldots),$$

where the sum is over all i in Y.

Composition Theorem   The cycle index sum $Z(\mathscr{F})$ is obtained by composing $Z(A)$ around $\sum Z(B_i)$:

$$Z(\mathscr{F}) = Z(A)[\textstyle\sum Z(B_i)], \tag{8.3.2}$$

where the sum is over all i in Y.

Proof   We consider the representation of A as the power group E^A with object set Y^X; that is, $A' = E^A$ and for each α in A, there exists α' in A' such that for any f in Y^X

$$(\alpha'f)x = f(\alpha x), \tag{8.3.3}$$

for all x in X.

Let R be the ring of power series in the variables $s_1, s_2, s_3, \ldots$ over the rationals. We define the map $\varphi : A' \times Y^X \to R$ by

$$\varphi(\alpha'; f) = \prod_k \prod_{i \in f(X)} Z(B_i; s_k, s_{2k}, \ldots)^{j_k(\alpha, X_i)}. \tag{8.3.4}$$

One can check that φ satisfies condition (8.1.1). Then on applying Lemma (8.1.2) above we have

$$|A|^{-1} \sum_{\alpha \in A} \sum_{f = \alpha' f} \varphi(\alpha', f) = \sum_F |A(f)|^{-1} \sum_{\alpha \in A(f)} \varphi(\alpha', f), \tag{8.3.5}$$

where the sum on the right side of (8.3.5) is over all classes F in $\mathscr{F}$ and f is some element of F. Note that although the lemma is only stated for the case in which the sums are finite, it still holds when all the sums are defined. The sums *are* defined in (8.3.5) because we have required that only finitely many B_i have the same degree.

From definition (8.3.4) for φ, and equation (8.2.2) for the cycle index of the composition group, it can be seen that the right side of (8.3.5) is $Z(\mathscr{F})$. It is also easily verified that for each α in A

$$\sum_{f = \alpha' f} \varphi(\alpha', f) = \prod_k \left(\sum_{i \in Y} Z(B_i; s_k, s_{2k}, \ldots) \right)^{j_k(\alpha)}, \tag{8.3.6}$$

and the proof is completed. $//$

8.4   CONNECTED GRAPHS

Our object is to determine the cycle index sum for all connected graphs as a function of the cycle index sum for all graphs. This can be accomplished with the aid of the Composition Theorem. If $\mathcal{G}$ is the set of all graphs, then from formula (7.2.18) we have

$$Z(\mathcal{G}) = \sum_{n=1}^{\infty} Z(S_n^{(1,2)}; s_k, 2), \tag{8.4.1}$$

and therefore we can regard $Z(\mathcal{G})$ as known. Now we shall take an alternative approach toward finding $Z(\mathcal{G})$, just as in the case of finding the ordinary generating function for connected graphs. Let $\mathcal{C}$ be the set of all connected graphs, and let $\mathcal{G}_n$ consist of the graphs which have exactly n components. With the symmetric group S_n acting on X, we consider all functions $f : X \to \mathcal{C}$. Each function represents that element of $\mathcal{G}_n$ which has $|f^{-1}(G)|$ components isomorphic to the connected graph G in $\mathcal{C}$. It is clear that functions which are equivalent with respect to S_n represent the same graph in $\mathcal{G}_n$. Since each graph in $\mathcal{C}$ has an automorphism group, each function f has associated with it a generalized composition group which is precisely the group of the graph represented by f. To obtain the cycle index sum $Z(\mathcal{G}_n)$ for all inequivalent functions, we may apply the Composition Theorem to obtain the following identity:

$$Z(\mathcal{G}_n) = Z(S_n)[Z(\mathcal{C})]. \tag{8.4.2}$$

On summing (8.4.2) over all n we have

$$Z(\mathcal{G}) = \sum_{n=1}^{\infty} Z(S_n)[Z(\mathcal{C})]. \tag{8.4.3}$$

But since

$$\sum_{n=0}^{\infty} Z(S_n) = \exp\left\{ \sum_{k=1}^{\infty} s_k/k \right\}, \tag{8.4.4}$$

(compare formula (3.1.1)), we have

$$1 + Z(\mathcal{G}) = \exp \sum_{k=1}^{\infty} (s_k/k)[Z(\mathcal{C})]. \tag{8.4.5}$$

From an idea by Cadogan [C1], (8.4.5) can be solved for $Z(\mathcal{C})$ using möbius inversion (see Robinson [R19] for details), and hence the theorem takes the following form.

Theorem If $\mathcal{G}$ and $\mathcal{C}$ are the sets of all graphs and connected graphs respectively, then

$$Z(\mathcal{C}) = \sum_{i=1}^{\infty} \frac{\mu(i)}{i} \sum_{j=1}^{\infty} \frac{(-1)^{j+1}}{j} s_i^j [Z(\mathcal{G})]. \tag{8.4.6}$$

From (8.4.1) one can obtain as the first few terms

$$Z(\mathcal{G}) = s_1 + s_1^2 + s_2 + \tfrac{4}{3}s_1^3 + 2s_1s_2 + \tfrac{2}{3}s_3 + \tfrac{3}{3}s_1^4$$

$$+ 4s_1^2s_2 + 2s_2^2 + \tfrac{4}{3}s_1s_3 + s_4 + \cdots. \qquad (8.4.7)$$

On substitution in (8.4.6) we have

$$Z(\mathcal{C}) = s_1 + \tfrac{1}{2}s_1^2 + \tfrac{1}{2}s_2 + \tfrac{2}{3}s_1^3 + s_1s_2 + \tfrac{1}{3}s_3$$

$$+ \tfrac{19}{12}s_1^4 + 2s_1^2s_2 + \tfrac{5}{4}s_2^2 + \tfrac{2}{3}s_1s_3 + \tfrac{1}{2}s_4 + \cdots. \qquad (8.4.8)$$

Note that on setting $s_k = 1$ in those parts of $Z(\mathcal{G})$ and $Z(\mathcal{C})$ which consist of terms of degree ≤ 4, one obtains the number 18 of graphs with ≤ 4 points and the number 10 of connected graphs with ≤ 4 points. It should also be emphasized that although formula (8.4.6) is an explicit expression for $Z(\mathcal{C})$, it is much easier to compute $Z(\mathcal{C})$ directly from (8.4.5). Specifically let $\mathcal{C}_p$ be the set of connected graphs on p points, and $\mathcal{G}(p)$ the set of graphs, each component of which contains at least p points. Then $\mathcal{G}(1) = \mathcal{G}$ and $Z(\mathcal{C}_p)$ consists precisely of the terms of order p in $Z(\mathcal{G}(p))$. From the Composition Theorem we have

$$1 + Z(\mathcal{G}(p + 1)) = (1 + Z(\mathcal{G}))\exp \sum_{k=1}^{\infty} -(s_k/k)[Z(\mathcal{C}_p)]. \qquad (8.4.9)$$

Therefore $Z(\mathcal{C}_p)$ is determined inductively for each p.

8.5 CYCLE INDEX SUMS FOR ROOTED GRAPHS

We now establish a relationship between cycle index sums for a collection of graphs and the corresponding rooted graphs. Since a graph G can be rooted at any of its points, a rooted graph may be considered to be an ordered pair (G, u) where u is a point of G. Two rooted graphs (G, u) and (G, v) are then the same (isomorphic) if and only if u and v are in the same orbit of the group of G. The automorphisms of (G, u) consist of all the automorphisms of G which fix u. For convenience, however, we do not include the root point in the object set of the group of any rooted graph. Suppose $\mathcal{H}$ is any set of graphs and $\mathcal{H}'$ consists of all the different rooted graphs which can be obtained by rooting the graphs in $\mathcal{H}$. By an abuse of notation we again use $Z(\mathcal{H})$ and $Z(\mathcal{H}')$ to denote the cycle index sums of all the graphs in $\mathcal{H}$ and the rooted graphs in $\mathcal{H}'$. Now Redfield's Lemma (7.2.3) may be applied to obtain a relationship between $Z(\mathcal{H})$ and $Z(\mathcal{H}')$.

Theorem The cycle index sum $Z(\mathcal{H}')$ for rooted graphs is the partial derivative of $Z(\mathcal{H})$ with respect to the first variable:

$$Z(\mathcal{H}') = \partial Z(\mathcal{H})/\partial s_1. \tag{8.5.1}$$

Proof It is obviously sufficient to prove the result when $\mathcal{H}$ consists of exactly one graph G. We apply Redfield's Lemma (7.2.3) to the group of G with the class function

$$\varphi(\alpha) = \prod s_k^{jk(\alpha)} \tag{8.5.2}$$

for each α in the group of G. Then the left side of (7.2.3) is $s_1 \, \partial Z(G)/\partial s_1$ and the right side is $s_1 Z(\mathcal{H}')$.   //

As an example, consider the complete graph K_n, whose group is S_n. This graph can be rooted in only one way and the group of the rooted graph is S_{n-1}. Therefore, we have

$$\partial Z(S_n)/\partial s_1 = Z(S_{n-1}), \tag{8.5.3}$$

which can also be verified by routine computation from (2.2.5).

8.6 BLOCKS

First we shall determine the cycle index sum for all rooted blocks. That is, if $\mathcal{B}$ is the set of all blocks, we seek an expression for $Z(\mathcal{B}')$. This is accomplished by finding $Z(\mathcal{C}')$ in terms of $Z(\mathcal{B}')$ and then inverting the relationship. We observe from the Composition Theorem that the cycle index sum for all connected graphs which are rooted at a point which is *not* a cutpoint is

$$Z(\mathcal{B}')[s_1 Z(\mathcal{C}')].$$

Again applying the Composition Theorem, it can be seen that the cycle index sum for all rooted, connected graphs in which the root is incident with exactly n blocks is

$$Z(S_n)[Z(\mathcal{B}')[s_1 Z(\mathcal{C}')]].$$

On summing over all n and applying the identity (8.4.4), we have

$$Z(\mathcal{C}') = \exp \sum_{k=1}^{\infty} (s_k/k)[Z(\mathcal{B}')[s_1 Z(\mathcal{C}')]]. \tag{8.6.1}$$

Since $z(\mathcal{C}') = \partial Z(\mathcal{C})/\partial s_1$, formula (8.6.1) could be inverted to find $Z(\mathcal{B}')$, but as in the determination of $Z(\mathcal{C})$ it is easier to use the exponential relation (8.6.1) directly as follows.

We let $\mathscr{C}'(p)$ be the set of rooted, connected graphs such that every block containing the root point has at least p points. Since the one-point graph has *no* blocks, we have

$$Z(\mathscr{C}'(2)) = Z(\mathscr{C}'). \tag{8.6.2}$$

If $\mathscr{B}_p$ is the set of all blocks with exactly p points, then $Z(\mathscr{B}'_p)$ consists of all the terms of order $p - 1$ in $Z(\mathscr{C}'(p))$.

Now let $\mathscr{H}'_p$ be the set of all rooted connected graphs with exactly p points in each block that contains the root. In the same manner as (8.6.1) was obtained we have

$$Z(\mathscr{H}'_p) = \exp \sum_{k=1}^{\infty} (s_k/k)[Z(\mathscr{B}'_p)[s_1 Z(\mathscr{C}')]]. \tag{8.6.3}$$

It is clear that

$$Z(\mathscr{C}'(p + 1))Z(\mathscr{H}'_p) = Z(\mathscr{C}'(p)), \tag{8.6.4}$$

and so combining (8.6.3) and (8.6.4) we have

$$Z(\mathscr{C}'(p + 1)) = Z(\mathscr{C}'(p)) \exp - \sum_{k=1}^{\infty} (s_k/k)\{[Z(\mathscr{B}'_p)[s_1 Z(\mathscr{C}')]]\}. \tag{8.6.5}$$

Therefore, if $Z(\mathscr{B}'_p)$ and $Z(\mathscr{C}'(p))$ are known, (8.6.5) can be used to obtain $Z(\mathscr{C}'(p + 1))$. But then $Z(\mathscr{B}'_{p+1})$ consists of all the terms of order p in $Z(\mathscr{C}'(p + 1))$ and hence $Z(\mathscr{B}'_p)$ is determined inductively. Finally.

$$Z(\mathscr{B}') = \sum_{p=2}^{\infty} Z(\mathscr{B}'_p), \tag{8.6.6}$$

and the sum starts at $p = 2$ because of our convention which excludes the one point graph from $\mathscr{B}$ and $\mathscr{B}'$.

Let $Z(\mathscr{B}_p)|_{s_1 = 0}$ be the series obtained from $Z(\mathscr{B}_p)$ on replacing each s_1 by zero; in other words $Z(\mathscr{B}_p)|_{s_1 = 0}$ is the contribution to $Z(\mathscr{B}_p)$ of all automorphisms which have no fixed points. At this point we observe that

$$Z(\mathscr{B}_p) = \int_0^{s_1} Z(\mathscr{B}'_p)\, ds_1 + Z(\mathscr{B}_p)|_{s_1 = 0}. \tag{8.6.7}$$

Therefore our remaining task is to obtain $Z(\mathscr{B}_p)|_{s_1 = 0}$. To this end we state the following result from [R19].

Theorem Every connected graph G which has a fixed-point free automorphism has a unique block whose points are permuted among themselves by all automorphisms of G.

We also have

$$Z(\mathscr{C}) = \int_0^{s_1} Z(\mathscr{C}') \, ds_1 + Z(\mathscr{C})|_{s_1 = 0} \qquad (8.6.8)$$

and from the Composition Theorem we know that the cycle index sum for all connected graphs rooted at a block is $Z(\mathscr{B})[s_1 Z(\mathscr{C}')]$.

Therefore by the theorem (8.5.1), the fixed-point free contribution to $Z(\mathscr{C})$ is just $Z(\mathscr{B})[s_1 Z(\mathscr{C}')]$ evaluated at $s_1 = 0$:

$$Z(\mathscr{C})|_{s_1 = 0} = Z(\mathscr{B})[s_1 Z(\mathscr{C}')]|_{s_1 = 0}. \qquad (8.6.9)$$

In practice, of course, we may use the identity:

$$(Z(\mathscr{B})|_{s_1 = 0})[s_1 Z(\mathscr{C}')] = Z(\mathscr{B})[s_1 Z(\mathscr{C}')]|_{s_1 = 0}. \qquad (8.6.10)$$

In order to determine $Z(\mathscr{B}_p)|_{s_1 = 0}$ inductively, let $\mathscr{B}(p)$ be the set of all non-separable graphs (blocks) with at least p points. Note that $\mathscr{B}(2) = \mathscr{B}$ and that $Z(\mathscr{B}_p)|_{s_1 = 0}$ consists of all the terms of order p in $Z(\mathscr{B}(p))[s_1 Z(\mathscr{C}')]|_{s_1 = 0}$. Since $\mathscr{B}(2) = \mathscr{B}$, (8.6.9) gives us

$$Z(\mathscr{B}(2))[s_1 Z(\mathscr{C}')]|_{s_1 = 0} = Z(\mathscr{C})|_{s_1 = 0}, \qquad (8.6.11)$$

and, of course,

$$Z(\mathscr{B}_2)|_{s_1 = 0} = s_2/2. \qquad (8.6.12)$$

The next question, which follows from the fact that $\mathscr{B}(p)$ is the disjoint union of $\mathscr{B}_p$ and $\mathscr{B}(p + 1)$, completes the induction argument:

$$Z(\mathscr{B}(p + 1))[s_1 Z(\mathscr{C}')]|_{s_1 = 0} = Z(\mathscr{B}(p))[s_1 Z(\mathscr{C}')]|_{s_1 = 0}$$
$$- Z(\mathscr{B}_p)[s_1 Z(\mathscr{C}')]|_{s_1 = 0}. \qquad (8.6.13)$$

Specifically, if $Z(\mathscr{B}_p)|_{s_1 = 0}$ and $Z(\mathscr{B}(p))[s_1 Z(\mathscr{C}')]|_{s_1 = 0}$ are known, then (8.6.13) determines $Z(\mathscr{B}(p + 1))[s_1 Z(\mathscr{C}')]|_{s_1 = 0}$, whose terms of order $p + 1$ are precisely the terms of $Z(\mathscr{B}_{p+1})|_{s_1 = 0}$.

Finally, we have

$$Z(\mathscr{B})|_{s_1 = 0} = \sum_{p=2}^{\infty} Z(\mathscr{B}_p)|_{s_1 = 0} \qquad (8.6.14)$$

and Robinson's Enumeration Theorem for blocks may be stated in the following way.

Theorem The cycle index sum $Z(\mathscr{B})$ for blocks is given by

$$Z(\mathscr{B}) = \int_0^{s_1} Z(\mathscr{B}') \, ds_1 + Z(\mathscr{B})|_{s_1 = 0} \qquad (8.6.15)$$

where $Z(\mathscr{B}')$ is determined by (8.6.1) and $Z(\mathscr{B})|_{s_1 = 0}$ by (8.6.11) and (8.6.13).

TABLE 8.6.1

THE NUMBER OF BLOCKS OF ORDER p

p	1	2	3	4	5	6	7	8	9
b_p	0	1	1	3	10	56	468	7 123	194 066

The reader may find it instructive to use this method to determine the terms of order less than five in $Z(\mathscr{B})$:

$$Z(\mathscr{B}) = \tfrac{1}{2}s_1^2 + \tfrac{1}{2}s_2 + \tfrac{1}{6}s_1^3 + \tfrac{1}{2}s_1s_2 + \tfrac{1}{3}s_3 + \tfrac{5}{12}s_1^4 + s_1^2s_2$$
$$+ \tfrac{3}{4}s_2^2 + \tfrac{1}{3}s_1s_3 + \tfrac{1}{2}s_4 + \cdots. \tag{8.6.16}$$

The number of blocks of order p is, of course, just the coefficient sum of $Z(\mathscr{B}_p)$.

This method has been programmed by Osterweil [O3] for a computer and the results for graphs with less than ten points are displayed in Table 8.6.1.

8.7 GRAPHS WITH GIVEN BLOCKS

Numerous classes of graphs can be defined in terms of the blocks which are contained in the members of the class. For example, trees are the connected graphs whose blocks are all isomorphic to K_2; block graphs have *all* of their blocks complete; cacti are the connected graphs whose blocks are lines or cycles; connected graphs with no points of degree ≤ 1 are precisely the nontrivial connected graphs none of whose end blocks are isomorphic to K_2, i.e., with no endpoints. To enumerate some of these classes of graphs we shall provide generalizations of some of the results of the previous section. We shall also provide a generalized form of the method devised by Norman [N1] for counting graphs with given blocks. The discussion again follows Robinson [R19]!

Let $\mathscr{D}$ be a class of blocks and let $\mathscr{K}$ be the set of connected graphs each of whose blocks is in $\mathscr{D}$ The first result, which expresses $Z(\mathscr{K}')$ in terms of $Z(\mathscr{D}')$ can be derived in a manner similar to the verification of (8.6.1).

Theorem   The cycle index sum $Z(\mathscr{K}')$ for rooted, connected graphs whose blocks are in the set $\mathscr{D}$ is given by

$$Z(\mathscr{K}') = \exp \sum_{k=1}^{\infty} (s_k/k)[Z(\mathscr{D}')[s_1 Z(\mathscr{K}')]]. \tag{8.7.1}$$

If $Z(\mathcal{D})$ is given, then $Z(\mathcal{D}') = \partial Z(\mathcal{D})/\partial s_1$ and formula (8.7.1) can be used to obtain $Z(\mathcal{K}')$. The next theorem expresses $Z(\mathcal{K})$ in terms of $Z(\mathcal{K}')$.

Theorem The cycle index sum $Z(\mathcal{K})$ for connected graphs whose blocks are in the set $\mathcal{D}$ is given by

$$Z(\mathcal{K}) = \int_0^{s_1} Z(\mathcal{K}') \, ds_1 + Z(\mathcal{D})|_{s_1 = 0}[s_1 Z(\mathcal{K}')]. \tag{8.7.2}$$

The proof of this result is similar to that of (8.6.9). Note that we could have used this theorem to establish formula (8.6.9), because with $\mathcal{K} = \mathcal{C}$ and $\mathcal{D} = \mathcal{B}$, formulas (8.7.2) implies (8.6.9). To further illustrate the use of (8.7.2), we shall show how the cycle index sum for trees can be derived. Let $\mathcal{T}$ be the set of all trees and let $\mathcal{D}$ consist only of the complete graph K_2. Thus $Z(\mathcal{D}') = s_1$ and by (8.7.1) we have

$$Z(\mathcal{T}') = \exp \sum_{k=1}^{\infty} (s_k/k)[s_1 Z(\mathcal{T}')]. \tag{8.7.3}$$

Now $Z(\mathcal{T}')$ can be determined recursively from this relation. For example, the terms of order less than 5 in $Z(\mathcal{T}')$ are

$$Z(\mathcal{T}') = 1 + s_1 + \tfrac{3}{2}s_1^2 + \tfrac{1}{2}s_2 + \tfrac{8}{3}s_1^3 + s_1 s_3 + \tfrac{1}{3}s_3$$
$$+ \tfrac{125}{24}s_1^4 + \tfrac{9}{4}s_1^2 s_2 + \tfrac{5}{8}s_2^2 + \tfrac{2}{3}s_1 s_3 + \tfrac{1}{4}s_4 + \cdots. \tag{8.7.4}$$

These terms correspond to the rooted trees of order less than 6. From (8.7.3) and (3.1.1) it follows that

$$Z(\mathcal{T}') = 1 + \sum_{k=1}^{5} Z(S_k)[s_1 Z(\mathcal{T}')] + \cdots. \tag{8.7.5}$$

Then the terms of order 5 in $Z(\mathcal{T}')$ may be computed in five contributions:

$$Z(S_1)[s_1 Z(\mathcal{T}')] = \cdots + \tfrac{125}{24}s_1^5 + \tfrac{9}{4}s_1^3 s_2$$
$$+ \tfrac{5}{8}s_1 s_2^2 + \tfrac{2}{3}s_1^2 s_3 + \tfrac{1}{4}s_1 s_4 + \cdots$$
$$Z(S_2)[s_1 Z(\mathcal{T}')] = \cdots + \tfrac{25}{6}s_1^5 + \tfrac{3}{2}s_1^3 s_2 + \tfrac{1}{2}s_1^2 s_3 + \cdots$$
$$Z(S_3)[s_1 Z(\mathcal{T}')] = \cdots + \tfrac{5}{4}s_1^5 + s_1^3 s_2 + \tfrac{3}{4}s_1 s_2^2 + \cdots \tag{8.7.6}$$
$$Z(S_4)[s_1 Z(\mathcal{T}')] = \cdots + \tfrac{1}{6}s_1^5 + \tfrac{1}{2}s_1^3 s_2 + \tfrac{1}{3}s_1^2 s_3 + \cdots$$
$$Z(S_5)[s_1 Z(\mathcal{T}')] = \tfrac{1}{120}s_1^5 + \tfrac{1}{12}s_1^3 s_2 + \tfrac{1}{8}s_1 s_2^2$$
$$+ \tfrac{1}{6}s_1^2 s_3 + \tfrac{1}{4}s_1 s_4 + \tfrac{1}{6}s_2 s_3 + \tfrac{1}{5}s_5 + \cdots.$$

Combining all terms of order 5 in $Z(\mathcal{T}')$, we have

$$\tfrac{54}{5}s_1^5 + \tfrac{16}{3}s_1^3 s_2 + \tfrac{3}{2}s_1 s_2^2 + \tfrac{3}{2}s_1^2 s_3 + \tfrac{1}{2}s_1 s_4 + \tfrac{1}{6}s_2 s_3 + \tfrac{1}{5}s_5.$$

Now $Z(\mathcal{D})|_{s_1=0} = \tfrac{1}{2}(s_1^2 + s_2)|_{s_1=0} = s_2/2$, and therefore from (8.7.2)

$$Z(\mathcal{T}) = \int_0^{s_1} Z(\mathcal{T}')\,ds_1 + (s_2/2)[s_1 Z(\mathcal{T}')]. \tag{8.7.7}$$

With (8.7.4) we can calculate

$$\tfrac{1}{2}s_2[s_1 Z(\mathcal{T}')] = \tfrac{1}{2}s_2 + \tfrac{3}{4}s_2^3 + \tfrac{1}{4}s_2 s_4 + \cdots, \tag{8.7.8}$$

and on integrating $Z(\mathcal{T}')$ and adding (8.7.8), we have the terms of order 6 in $Z(\mathcal{T})$:

$$\begin{aligned}
Z(\mathcal{T}) = {}& s_1 + \tfrac{1}{2}s_1^2 + \tfrac{1}{2}s_2 + \tfrac{1}{2}s_1^3 + \tfrac{1}{2}s_1 s_2 + \tfrac{2}{3}s_1^4 + \tfrac{1}{2}s_1^2 s_2 \\
& + \tfrac{1}{2}s_2^2 + \tfrac{1}{3}s_1 s_3 + \tfrac{25}{24}s_1^5 + \tfrac{3}{4}s_1^3 s_2 + \tfrac{5}{8}s_1 s_2^2 \\
& + \tfrac{1}{3}s_1^2 s_3 + \tfrac{1}{4}s_1 s_4 + \tfrac{9}{5}s_1^6 + \tfrac{4}{3}s_1^4 s_2 + \tfrac{3}{4}s_1^2 s_2^2 \\
& + \tfrac{1}{2}s_1^3 s_3 + \tfrac{3}{4}s_2^3 + \tfrac{1}{4}s_1^2 s_4 + \tfrac{1}{4}s_2 s_4 + \tfrac{1}{6}s_1 s_2 s_3 \\
& + \tfrac{1}{5}s_1 s_5 + \cdots. \tag{8.7.9}
\end{aligned}$$

This result is checked by summing the coefficients and comparing the number with the number of trees on fewer than seven points (see equation (3.2.8)).

Next we introduce Robinson's generalization of Norman's method [N1] for determining the number of graphs with given blocks.

Theorem   The cycle index sum $Z(\mathcal{K})$ for graphs whose blocks are in the set $\mathcal{D}$ is given by

$$Z(\mathcal{K}) = (s_1 + Z(\mathcal{D}) - s_1 Z(\mathcal{D}'))[s_1 Z(\mathcal{K})]. \tag{8.7.10}$$

Note that in (8.7.1) and (8.7.10) $Z(\mathcal{K})$ and $Z(\mathcal{K}')$ can be replaced by ordinary generating functions, but this cannot be done in (8.7.2). Furthermore, when cycle index sums are required, (8.7.2) is much easier to use than (8.7.10).

We conclude this section by using the techniques developed here to enumerate connected graphs without endpoints. The following theorem of Robinson [R19] enables us to obtain the appropriate cycle index sum after inversion.

Theorem The cycle index sum $Z(\mathcal{M})$ for the set $\mathcal{M}$ of all connected graphs with no endpoints satisfies

$$Z(\mathscr{C}) = Z(\mathscr{T}) + Z(\mathcal{M})[s_1 Z(\mathscr{T}')].\qquad (8.7.11)$$

The proof uses the Composition Theorem and the fact that connected graphs which are not trees correspond to the graphs to be enumerated with rooted trees attached to them. The latter correspondence also serves to provide the basis for counting the number of labeled graphs derived from $\mathcal{M}$.

To invert (8.7.11) note than an element of $\mathscr{C}$ on $p + 1$ points is either in $\mathcal{M}$ or its maximal subgraph which has no endpoints has less than $p + 1$ points. Now if $\mathcal{M}_p$ consists of those elements of $\mathcal{M}$ on at most p points, then the terms of order $p + 1$ in $Z(\mathcal{M}_p)[s_1 Z(\mathscr{T}')]$ correspond to all connected graphs on $p + 1$ points, except for trees, which *do* have endpoints. Therefore the terms of order $p + 1$ in $Z(\mathscr{T}) + Z(\mathcal{M}_p)[s_1 Z(\mathscr{T}')]$ are contributed by *all* connected graphs with $p + 1$ points and with at least one endpoint. Therefore $Z(\mathcal{M}_{p+1}) - Z(\mathcal{M}_p)$ consists of the terms of order $p + 1$ in

$$Z(\mathscr{C}) - (Z(\mathscr{T}) + Z(\mathcal{M}_p)[s_1 Z(\mathscr{T}')]).$$

For example, to determine $Z(\mathcal{M}_4) - Z(\mathcal{M}_3)$ we begin with

$$Z(\mathcal{M}_3) = \tfrac{1}{6}s_1^3 + \tfrac{1}{2}s_1 s_2 + \tfrac{1}{3}s_3 \qquad (8.7.12)$$

and from (8.7.4) and (8.7.12) we have

$$Z(\mathcal{M}_3)[s_1 Z(\mathscr{T}')] = \tfrac{1}{6}s_1^3 + \tfrac{1}{2}s_1 s_2 + \tfrac{1}{3}s_3 + \tfrac{1}{2}s_1^4 + \tfrac{1}{2}s_1^2 s_2 + \cdots. \quad (8.7.13)$$

Now from (8.4.8) for $Z(\mathscr{C})$, (8.7.4) and (8.7.13) we have

$$Z(\mathcal{M}_4) - Z(\mathcal{M}_3) = \tfrac{5}{12}s_1^4 + s_1^2 s_2 + \tfrac{3}{4}s_2^2 + \tfrac{1}{3}s_1 s_3 + \tfrac{1}{2}s_4. \qquad (8.7.14)$$

The coefficient sum in (8.7.14) is 3 and there are, indeed, exactly three connected graphs on four points which have no endpoints.

8.8 ACYCLIC DIGRAPHS

We have seen in Section 1.6 that every acyclic digraph has at least one point of indegree zero and that any extension of an acyclic digraph is also acyclic. Robinson [R20] used these facts to count both labeled and unlabeled acyclic digraphs. However, to count unlabeled acyclic digraphs, he found it necessary to incorporate information about the symmetries of all the acyclic digraphs. This was accomplished by using cycle index sums in which cycles of points of indegree zero are distinguished from the other cycles. Then a bilinear operation for cycle indexes can be defined so that the cycle index sum for extensions of any acyclic digraph D can be expressed in terms of $Z(D)$ and $Z(S_n)$. This leads to a recursive formula for the cycle index sum of all acyclic digraphs of order p.

We have expressed cycle indexes with two sets of variables in several places. To distinguish between points and lines, s_k and t_k were used in section 4.4 for counting supergraphs and in formula (7.2.19) for the cycle index sum of the point-line groups of graphs of order p. To count mixed graphs in Section 5.4, s_k and t_k appear in the cycle index of the reduced ordered pair group (5.4.5) to stand for pairs of converse cycles and self-converse cycles respectively.

Here we express the cycle index of the group of a digraph D using two sets of variables: s_k for cycles of points of indegree zero and t_k for cycles of the other points. With this convention, the acyclic digraphs D_1 and D_2 of Figure 1.6.2 have $Z(D_1) = s_1^2 t_1$ and $Z(D_2) = \frac{1}{2}(s_1^2 t_1 + s_2 t_1)$.

Now Redfield's Lemma (7.2.3) can be used to obtain an alternative expression for the cycle index sum of all the extensions of order p of an acyclic digraph D_0 of order $p - n$. Let $A = S_n \times \Gamma(D_0)$ and note that A induces a group A' whose object set consists of all extensions of D_0 to acyclic digraphs with n points of indegree zero. This representation of A can be made explicit by using a restricted power group, but since we have used this approach so often, a hint should be sufficient. For each β in $\Gamma(D_0)$, let j_k be the number of cycles of length k of points of indegree zero, and let m_k be the number of cycles of length k of the other points of D_0. For α in S_n, let i_k be the number of cycles of length k. This looks like a surfeit of variables but their presence will make our formulas less cluttered. Then for each (α, β) in A we define the class function φ for A by

$$\varphi(\alpha, \beta) = \prod s_k^{i_k} t_k^{j_k + m_k} . \tag{8.8.1}$$

The permutation in A' induced by (α, β) is denoted by $(\alpha, \beta)'$ and $j_1((\alpha, \beta)')$ is the number of extensions fixed by $(\alpha, \beta)'$. On applying (7.2.3) we can conclude that

$$\sum Z(D) = n! |\Gamma(D_0)|^{-1} \sum_{(\alpha, \beta)} j_1((\alpha, \beta)') \varphi(\alpha, \beta), \tag{8.8.2}$$

where the first sum is over all extensions of D_0 with n points of indegree zero.

The number $j_1((\alpha, \beta)')$ depends only on the cycle structure of α and β and can be determined explicitly:

$$j_1((\alpha, \beta)') = \prod_r (2^{\Sigma_k (k,r) i_k} - 1)^{j_r} 2^{\Sigma_k (k,r) i_k m_r}. \tag{8.8.3}$$

Note the resemblance of the right side of (8.8.3) to the contribution

$$\prod_{r,k} s_{[r,k]}^{(r,k)(j_r + m_r) i_k}$$

of (α, β) to the cycle index of the cartesian product $S_n \times \Gamma(D_0)$ as in (4.3.10). Replacing each variable $s_{[r,k]}$ in this expression by 2 gives the right side of (8.8.3) except for the term -1. The presence of the latter ensures that in each digraph fixed by $(\alpha, \beta)'$ *all* the points of D_0 have positive indegree.

In order to express the counting theorem entirely in terms of cycle indexes, the bilinear operation $*$ is defined for monomials by

$$\prod s_k^{i_k} * \prod_j s_j^{j_k} t_k^m = j_1((\alpha, \beta)')\varphi(\alpha, \beta) \tag{8.8.4}$$

where the right side is determined by (8.8.1) and (8.8.3). On extending this operation linearly, we can restate (8.8.2) as

$$\sum Z(D) = Z(S_n) * Z(D_0). \tag{8.8.5}$$

Finally on summing this equation we have the main result.

Theorem   The cycle index sum $Z(\mathcal{H}_p)$ for the set $\mathcal{H}_p$ of acyclic digraphs of order p satisfies

$$Z(\mathcal{H}_p) = \sum_{n=1}^{p} Z(S_n) * Z(\mathcal{H}_{p-n}) \tag{8.8.6}$$

where $Z(\mathcal{H}_0) = 1$ by definition.

Robinson chose to express his counting formula in terms of the cycle index sum for the entire set $\mathcal{H}$ of acyclic digraphs. On summing (8.8.6) over all p, his equation is obtained

$$Z(\mathcal{H}) = \sum_{n=1}^{\infty} Z(S_n) * (1 + Z(\mathcal{H})). \tag{8.8.7}$$

Equation (8.8.6), however, is what one actually uses for calculating the number of a_p of acyclic digraphs of order p. For example, to determine a_4, we need $Z(\mathcal{H}_1)$, $Z(\mathcal{H}_2)$, and $Z(\mathcal{H}_3)$. These are easily found with the aid of Figure 8.8.1.

$Z(\mathcal{H}_1) = s_1$

$Z(\mathcal{H}_2) = \frac{1}{2}(s_1^2 + s_2) + s_1 t_1$

$$Z(\mathcal{H}_3) = \frac{1}{6}(s_1^3 + 3s_1 s_2 + 2s_3) + \frac{3}{2}s_1^2 t_1 + \frac{1}{2}s_1 t_2 + \frac{5}{2}s_1 t_1^2 + \frac{1}{2}s_2 t_1 \tag{8.8.8}$$

Figure 8.8.1

The acyclic digraphs of order 3.

TABLE 8.8.1

ACYCLIC DIGRAPHS

p	1	2	3	4	5	6
a_p	1	2	6	31	302	5 984

Using (8.8.1) and (8.8.3) as prescribed in the definition of the bilinear operation $*$ in (8.8.4), we have

$$Z(S_4) * Z(\mathcal{H}_0) = Z(S_4)$$

$$Z(S_3) * Z(\mathcal{H}_1) = \tfrac{7}{6}s_1^3 + \tfrac{3}{2}s_1 s_2 t_1 + \tfrac{1}{3}s_3 t_1$$

$$Z(S_2) * Z(\mathcal{H}_2) = \tfrac{33}{4}s_1^2 t_1^2 + \tfrac{3}{4}s_1^2 t_2 + \tfrac{5}{4}s_2 t_1^2 + \tfrac{3}{4}s_2 t_2$$

$$Z(S_1) * Z(\mathcal{H}_3) = \tfrac{79}{6}s_1 t_1^3 + \tfrac{5}{2}s_1 t_1 t_2 + \tfrac{1}{3}s_1 t_3. \tag{8.8.9}$$

On adding these four equations (8.8.9), we obtain $Z(\mathcal{H}_4)$, whose coefficient sum 31 is the number of acyclic digraphs of order 4. Robinson used this method to compute the numbers in Table 8.8.1.

EXERCISES

8.1   Find the cycle index sum of all graphs with four components which consist of triangles and cycles of order 4.

8.2   Find the cycle index sum $Z(\mathcal{B}_5)$ for all blocks of order 5 (see (8.6.16) for $Z(\mathcal{B}_k)$ with $k < 5$).

8.3   Find the first few terms of the generating functions which enumerate connected graphs whose blocks consist of complete graphs of order (a) 2, (b) 3, (c) 4, or (d) cycles of order 4.

8.4   Use the methods of this chapter to enumerate (a) cacti and (b) block graphs.

8.5   Draw the 31 acyclic digraphs of order 4 (see Table 8.8.1). (*Hint*:   Use Appendix 2 of Harary [H1].)

It is not the business of the botanist to eradicate the weeds. Enough for him if he can tell us just how fast they grow.

C. Northcote Parkinson

Chapter 9 | ASYMPTOTICS

There are basically two methods for determining asymptotic formulas for graphs, and both are essentially contained in the published and unpublished work of Pólya, in the form of a letter to us in 1951. The choice of method depends on whether or not the graphs, or structures to be enumerated, are "treelike."

Both Riddell [R14] and Pólya investigated the asymptotic numbers of graphs, and some of their results were improved by Oberschelp [O1] and Wright [W5, W6]. In the first three sections of this chapter we provide the details of methods for determining the asymptotic number of graphs and digraphs. It appears that these methods can be applied successfully to various classes of graphs; for example, see Palmer [P5] and Harary [H8].

In Section 9.4 we consider asymptotic estimates for connected graphs and blocks. The results of Section 9.1 can then be used to show that almost all graphs are connected and that almost all graphs are blocks.

In [P8], Pólya found asymptotic results for those trees which represent the saturated hydrocarbons: all points have degree one or four. His approach was improved and generalized by Otter [O4] and applied by Ford and Uhlenbeck in the series [FU1] to graphs with given blocks, and hence cacti, as well as ordinary trees. Section 9.5 presents Otter's treatment of this problem.

9.1   GRAPHS

Pólya obtained the following formula for the asymptotic number of graphs g_p of order p:

$$g_p \sim 2^{\binom{p}{2}}/p! \qquad (9.1.1)$$

It follows from formula (4.1.9) that $2^{\binom{p}{2}}/p!$ is the largest term contributed to g_p by $Z(S_p^{(2)}, 2)$. Heuristically, the dominance of this term may be explained by observing that dividing the number $2^{\binom{p}{2}}$ of labeled graphs by $p!$ "removes the labels." Formula (9.1.1) can be improved, see Oberschelp [O1], to obtain better approximations for g_p up to remainder terms of order $(2^{\binom{p}{2}}p^k)/(p!2^{pk/2})$ for $k = 2, 3, \ldots$. The next theorem includes the improvement with $k = 3$.

Theorem   The number g_p of graphs of order p satisfies

$$g_p = \frac{2^{\binom{p}{2}}}{p!}\left(1 + \frac{p^2 - p}{2^{p-1}} + \mathcal{O}\left(\frac{p^3}{2^{3p/2}}\right)\right). \qquad (9.1.2)$$

Proof   Recall that the number of permutations in S_p which correspond to the partition (j) of p is denoted by $h(j)$ and is specified in formula (2.2.4). From formula (4.1.9) for the cycle index of the pair group of S_p, the number $q(j)$ of line-cycles determined by each of these permutations is

$$q(j) = \sum_k \left[\frac{k}{2}\right]j_k + \sum_k k\binom{j_k}{2} + \sum_{r<t}(r, t)j_r j_t. \qquad (9.1.3)$$

Now for each $k = 0$ through p, we let $g_p^{(k)}$ be the contribution to g_p determined by all partitions (j) with exactly $p - k$ parts equal to 1; that is,

$$g_p^{(k)} = (1/p!)\sum h(j)2^{q(j)}, \qquad (9.1.4)$$

where the sum is over all partitions (j) with $j_1 = p - k$. Then, of course,

$$g_p = \sum_{k=0}^{p} g_p^{(k)}, \qquad (9.1.5)$$

and from formulas (9.1.3) and (9.1.4) it follows quickly that

$$g_p^{(0)} = 2^{\binom{p}{2}}/p!. \tag{9.1.6}$$

For $k = 2$, consider the partition $(j) = (p - 2, 1, 0, \ldots, 0)$ with $j_1 = p - 2$ and $j_2 = 1$. From (9.1.3) we have

$$q(p - 2, 1, 0, \ldots, 0) = (p^2 - 3p + 4)/2. \tag{9.1.7}$$

Therefore, from (9.1.4), we may express

$$g_p^{(2)}/g_p^{(0)} = 2p(p - 1)/2^p. \tag{9.1.8}$$

Similarly, we find that

$$g_p^{(3)}/g_p^{(0)} = 2^4 p(p - 1)(p - 2)/(3 \cdot 2^{2p}) \tag{9.1.9}$$

and

$$g_p^{(4)}/g_p^{(0)} = p(p - 1)(p - 2)(p - 3)(2^3/2^{2p} + 2^6/2^{3p}). \tag{9.1.10}$$

Lemma   For each positive integer n

$$g_p \sim \sum_{k=0}^{n-1} g_p^{(k)}. \tag{9.1.11}$$

Proof of Lemma   This is accomplished by first establishing upper bounds for $g_p^{(k)}$. For each k, we consider those partitions (j) of p with $j_1 = p - k$. On substituting $j_1 = p - k$ and $j_2 = k/2$ in the right side of (9.1.3), an upper bound is obtained for $q(j)$:

$$q(j) \le \binom{p}{2} + \frac{1}{2}(k - pk + k^2/2). \tag{9.1.12}$$

Furthermore, the number of permutations of p objects with exactly $(p - k)$ objects fixed is less than or equal to $p!/(p - k)!$ (see Riordan [R15, p. 59]). Therefore, we can write

$$g_p^{(k)} \le \frac{1}{p!} \cdot \frac{p!}{(p - k)!} 2^{\binom{p}{2} + (k - pk + k^2/2)/2}. \tag{9.1.13}$$

Since $p!/(p - k)!$ is bounded by p^k, the bound for $g_{p,k}$ may be increased to yield

$$g_p^{(k)} \le g_p^{(0)} p^k/(2^{(p-1-k/2)/2})^k. \tag{9.1.14}$$

We always have $k \le p$, and therefore

$$g_p^{(k)} \le g_p^{(0)}(p/(2^{(p/4-1/2)}))^k. \tag{9.1.15}$$

On summing from $k = n$ to p, we have

$$\sum_{k=n}^{p} g_p^{(k)} \le g_p^{(0)} \sum_{k=n}^{p} (\sqrt{2}p/2^{p/4})^k. \qquad (9.1.16)$$

But the sum on the right side of (9.1.16) is the partial sum of a geometric series whose common ratio approaches zero as p increases. Therefore

$$\sum_{k=n}^{p} g_p^{(k)} \le c g_p^{(0)} (\sqrt{2}p/2^{p/4})^n \qquad (9.1.17)$$

where $c > 1$ is close to 1 for large p.

In particular, since

$$\sum_{k=0}^{n-1} g_p^{(k)} \le g_p \le \sum_{k=0}^{n-1} g_p^{(k)} + g_p^{(0)} \mathcal{O}(p^n/2^{pn/4}), \qquad (9.1.18)$$

we have on division

$$1 \le g_p / \sum_{k=0}^{n-1} g_p^{(k)} \le 1 + \mathcal{O}(p^n/2^{pn/4}), \qquad (9.1.19)$$

which verifies (9.1.11), and proves the lemma.

We can now continue with the proof of the theorem. To secure the error on the bound, we return to (9.1.14) and note that with $k = n$,

$$g_p^{(n)} \le 2^{(n/2 + n^2/4)} g_p^{(0)} p^n/2^{pn/2}, \qquad (9.1.20)$$

and hence

$$\sum_{k=n}^{2n} g_p^{(k)} = (g_p^{(0)}) \mathcal{O}(p^n/2^{pn/2}). \qquad (9.1.21)$$

From (9.1.17) it follows that

$$\sum_{k=2n+1}^{p} g_p^{(k)} = (g_p^{(0)}) \mathcal{O}((\sqrt{2}p/2^{p/4})^{2n+1}). \qquad (9.1.22)$$

But since

$$\mathcal{O}((\sqrt{2}p/2^{p/4})^{2n+1}) = \mathcal{O}(p^n/2^{pn/2}). \qquad (9.1.23)$$

We can combine (9.1.21) and (9.1.22) to find

$$\sum_{k=n}^{p} g_p^{(k)} = (g_p^{(0)}) \mathcal{O}(p^n/2^{pn/2}). \qquad (9.1.24)$$

The proof of the theorem (9.1.2) is completed by setting $n = 3$ in (9.1.24) and adding $g_p^{(0)} + g_p^{(2)}$ to both sides.   //

We conclude this section by improving the formula for g_p with $g_p^{(3)}$ and $g_p^{(4)}$. From (9.1.9), (9.1.10), and (9.1.24) we have

$$g_p = \frac{2^{\binom{p}{2}}}{p!}\left(1 + 2\frac{p^2 - p}{2^p} + 8\frac{p!}{(p-4)!}\frac{(3p - 7)/(3p - 9)}{2^{2p}} + \mathcal{O}\left(\frac{p^5}{2^{5p/2}}\right)\right).$$

(9.1.25)

Note that the second part of $g_p^{(4)}$ is incorporated in the error term. Formula (9.1.2) was used by Oberschelp [O1] to obtain the first and second approximations in Table 9.1.1 for g_p and (9.1.25) for the third approximations.

TABLE 9.1.1

APPROXIMATIONS FOR THE NUMBER OF GRAPHS

		Approximation		
p	g_p	First	Second	Third
2	2	1	2	2
3	4	1.333	3.333	4
4	11	2.667	6.667	10
5	34	8.533	19.20	29.87
6	156	45.51	88.18	127.3
7	1044	416.1	689.12	896.4
8	12 346	6 660	9 570	11 120

9.2 DIGRAPHS

The development of asymptotic formulas for digraphs is similar to that for graphs, and so we shall only sketch the proof of the theorem which provides the formula and the error bound. From formula (5.1.5) for the cycle index of the reduced ordered pair group of S_p, we see that the number $q(j)$ of line-cycles determined by each permutation which corresponds to the partition (j) is

$$q(j) = \sum_k (kj_k^2 - j_k) + 2\sum_{r<t}(r, t)j_r j_t.$$

(9.2.1)

As before, for each $k = 0$ through p we let

$$d_p^{(k)} = (1/p!)\sum h(j)2^{q(j)},$$

(9.2.2)

where the sum is over all partitions (j) with $j_1 = p - k$. Then the number d_p of digraphs of order p is the sum of the $d_p^{(k)}$ and in particular

$$d_p^{(0)} = 2^{p^2 - p}/p! \qquad (9.2.3)$$

$$d_p^{(2)}/d_p^{(0)} = 2^2 p(p - 1)/2^{2p} \qquad (9.2.4)$$

$$d_p^{(3)}/d_p^{(0)} = 2^8 p(p - 1)(p - 2)/(3 \cdot 2^{4p}) \qquad (9.2.5)$$

and

$$d_p^{(4)}/d_p^{(0)} = (p!/(p - 4)!)(2^7/2^{4p} + 2^{13}/2^{6p}). \qquad (9.2.6)$$

A bound on $q(j)$ is obtained by setting $j_1 = p - k$ and $j_2 = k/2$ in the right side of (9.2.1):

$$q(j) \leq p^2 - p + k(1 - 2p + k)/2 \qquad (9.2.7)$$

whenever $j_1 = p - k$.
Therefore, it follows that

$$d_p^{(k)} \leq d_p^{(0)}(p/2^{(2p-1-k)/2})^k. \qquad (9.2.8)$$

As in the case for graphs, it can now be shown that

$$d_p \sim \sum_{k=0}^{n-1} d_p^{(k)} \qquad (9.2.9)$$

and

$$\sum_{k=n}^{p} d_p^{(k)} = (d_p^{(0)})\mathcal{O}(p^n/2^{pn}). \qquad (9.2.10)$$

Hence we have the next theorem for digraphs.

Theorem The number d_p of digraphs of order p satisfies

$$d_p = \frac{2^{p^2 - p}}{p!}\left(1 + \frac{4p(p - 1)}{2^{2p}} + \mathcal{O}\left(\frac{p^3}{2^{3p}}\right)\right). \qquad (9.2.11)$$

On finding $d_p^{(3)}$ and $d_p^{(4)}$ our result takes the form

$$d_p = \frac{2^{p^2 - p}}{p!}\left(1 + \frac{4p(p - 1)}{2^{2p}} + 2^7 \frac{p!}{(p - 4)!} \frac{(3p - 7)/(3p - 9)}{2^{4p}} + \mathcal{O}\left(\frac{p^5}{2^{5p}}\right)\right).$$

$$(9.2.12)$$

9.3 GRAPHS WITH A GIVEN NUMBER OF POINTS AND LINES

We have seen that the term $2^{\binom{p}{2}}/p!$ contributed to $Z(S_p^{(2)}, 2)$ by the identity permutation is asymptotic to the total number of graphs of order p. Hence it would seem that to approximate the number $g_{p,q}$ of (p, q) graphs, we should try the contribution to the coefficient of x^k in $Z(S_p^{(2)}, 1 + x)$ made by the identity permutation. That this contribution is $\binom{\binom{p}{2}}{q}/p!$ can be seen when we express the graph counting polynomial as follows:

$$Z(S_p^{(2)}, 1 + x) = \frac{(1 + x)^{\binom{p}{2}}}{p!}(1 + f(x)), \tag{9.3.1}$$

with $f(x)$ defined by this equation. One would expect this approximation to be valid because $\binom{\binom{p}{2}}{q}/p!$ is just by (1.1.1) the number of labeled (p, q) graphs divided by $p!$ to "remove the labels." But we shall see that it can only be used when q is not near the ends of its range. For example, $g_{p,0} = 1$ for all p but $\binom{\binom{p}{2}}{0}/p! = 1/p!$ approaches zero as p increases. In fact, if q is a fixed constant then, of course

$$\binom{\binom{p}{2}}{q}\bigg/p! \sim 0. \tag{9.3.2}$$

Bounds on q for which the estimate of $g_{p,q}$ is valid were established by Pólya as mentioned in Ford and Uhlenbeck [FU1] and subsequently improved by Oberschelp [O1], as in the next theorem. Wright [W7] has recently found a necessary and sufficient condition for (9.3.4); see Exercise 9.13.

Theorem If for some ε with $0 < \varepsilon < \frac{1}{2}$,

$$\left|\frac{1}{2}\binom{p}{2} - q\right| = \mathcal{O}(p^{3/2 - \varepsilon}), \tag{9.3.3}$$

then the number $g_{p,q}$ of (p, q) graphs satisfies

$$g_{p,q} \sim \binom{\binom{p}{2}}{q}\bigg/p!. \tag{9.3.4}$$

Proof From the definition of $f(x)$ in (9.3.1), we have for all p and q

$$\binom{\binom{p}{2}}{q}\bigg/p! \leq g_{p,q} \leq \binom{\binom{p}{2}}{q}\bigg/p! + f(1)2^{\binom{p}{2}}/p!. \tag{9.3.5}$$

Hence it suffices to show that if q satisfies condition (9.3.3), then

$$f(1)2^{\binom{p}{2}}\bigg/\binom{\binom{p}{2}}{q} \sim 0. \tag{9.3.6}$$

But from the asymptotic behavior of g_p in (9.1.2) we observe that

$$f(1) = \mathcal{O}(p^2/2^p). \tag{9.3.7}$$

Therefore we need to show that for suitable values of q,

$$F(p, q) = (2^{\binom{p}{2} - p}p^2) \Bigg/ \binom{\binom{p}{2}}{q} \sim 0. \tag{9.3.8}$$

To do this we prove that there are positive constants c and δ with $0 \le \delta \le \frac{1}{2}$ such that for all p sufficiently large

$$\left| \frac{1}{2}\binom{p}{2} - q \right| < cp^{1+\delta} \tag{9.3.9}$$

implies (9.3.8). Then ε can be chosen so that (9.3.3) holds.

Now $F(p, q)$, the left side of (9.3.8), is largest when q is at the end of its range. Thus it follows from (9.3.9) that

$$F(p, q) \le \frac{(p^2 2^{(p^2 - 3p)/2})}{\left(\dfrac{\binom{p}{2}}{\frac{1}{2}\binom{p}{2} - cp^{1+\delta}} \right)}. \tag{9.3.10}$$

On applying Stirling's formula to the factorials in the denominator of the right side of (9.3.10) we have for some constant C_1 and p sufficiently large

$$F(p, q) \le C_1 \frac{(p^2 2^{(p^2 - 3p)/2})}{\binom{p}{2}^{\binom{p}{2} + 1/2}} a_p^{a_p} b_p^{b_p} (a_p b_p)^{1/2} \tag{9.3.11}$$

where

$$a_p = \frac{1}{2}\binom{p}{2} - cp^{1+\delta}, \tag{9.3.12}$$

and

$$b_p = \frac{1}{2}\binom{p}{2} + cn^{1+\delta}. \tag{9.3.13}$$

Since $(a_p b_p)^{1/2} = \mathcal{O}(p^2)$, we can simplify the bound on $F(p, q)$ further and write, for some constant C_2

$$F(p, q) \le C_2 \frac{p^4}{(2^p(p^2 - p)^{\binom{p}{2} + 1/2}} (p^2 - p - 4cp^{1+\delta})^{a_p} (p^2 - p + 4cp^{1+\delta})^{b_p}.$$

$$\tag{9.3.14}$$

On factoring and using the fact that $a_p + b_p = \binom{p}{2}$, we have

$$F(p, q) \leq C_2 \frac{p^4}{(2^p(p^2 - p)^{\binom{p}{2} + 1/2})}(p^2 - p)^{a_p}\left(1 - \frac{4cp^{1+\delta}}{p^2 - p}\right)^{a_p}$$

$$\times (p^2 - p)^{b_p}\left(1 + \frac{4cp^{1+\delta}}{p^2 - p}\right)^{b_p}$$

$$\leq C_2 \frac{p^4}{2^p(p^2 - p)^{1/2}}\left(1 - \frac{4cp^{1+\delta}}{p^2 - p}\right)^{a_p}\left(1 + \frac{4cp^{1+\delta}}{p^2 - p}\right)^{b_p}. \tag{9.3.15}$$

It suits our purpose to express the last inequality as follows

$$F(p, q) \leq C_2 \frac{p^4}{2^p(p^2 - p)^{1/2}}\left(\left(1 - \frac{cp^{1+\delta}}{\binom{p}{2}/2}\right)\left(1 + \frac{cp^{1+\delta}}{\binom{p}{2}/2}\right)\right)^{\binom{p}{2}/2}$$

$$\times \left(\left(1 + \frac{cp^{1+\delta}}{\binom{p}{2}/2}\right)\Big/\left(1 - \frac{cp^{1+\delta}}{\binom{p}{2}/2}\right)\right)^{cp^{1+\delta}}. \tag{9.3.16}$$

Next we use the identities

$$\left(\left(1 - \frac{cp^{1+\delta}}{\binom{p}{2}/2}\right)\left(1 + \frac{cp^{1+\delta}}{\binom{p}{2}/2}\right)\right)^{\binom{p}{2}/2} = \left(1 - \frac{c^2 p^{2+2\delta}/(\binom{p}{2}/2)}{\binom{p}{2}/2}\right)^{\binom{p}{2}/2} \tag{9.3.17}$$

and

$$\left|\frac{1 + \frac{cp^{1+\delta}}{\binom{p}{2}/2}}{1 - \frac{cp^{1+\delta}}{\binom{p}{2}/2}}\right|^{cp^{1+\delta}} = \left|\frac{1 + \frac{c^2 p^{2+2\delta}/(\binom{p}{2}/2)}{cp^{1+\delta}}}{1 - \frac{c^2 p^{2+2\delta}/(\binom{p}{2}/2)}{cp^{1+\delta}}}\right|^{cp^{1+\delta}}. \tag{9.3.18}$$

It can be shown that for any $x, y > 0$,

$$e^{xy/(y-x)} < (1 + x/y)^y < e^x \tag{9.3.19}$$

and furthermore for $y \geq x > 0$,

$$e^{-xy/(y-x)} < (1 - x/y)^y < e^{-x}. \tag{9.3.20}$$

On applying (9.3.19) and (9.3.20) to (9.3.17) and (9.3.18) we have

$$F(p, q) \leq C_2(p^4/(2^p(p^2 - p)^{1/2})\exp s(p) \tag{9.3.21}$$

where

$$s(p) = 4c^2 p^{2\delta}\Big/\left(1 - \frac{1 + 4cp^\delta}{p}\right). \tag{9.3.22}$$

If $\delta < 1$, we have for any constant $K > 4c^2$,

$$s(p) \le Kp^{2\delta} \tag{9.3.23}$$

provided p is large.

Thus we have for large p,

$$F(p, q) < C_2 p^4 e^{Kp^{2\delta}} / (2^p(p^2 - p))^{1/2} \tag{9.3.24}$$

where $K > 4c^2$ and $\delta < 1$.

If we choose $\delta \le \frac{1}{2}$, then

$$F(p, q) = \mathcal{O}(p^3)\left(\frac{e^K}{2}\right)^p. \tag{9.3.25}$$

Hence $F(p, q) \sim 0$ provided $K < \log 2$. For K to exist, then, the only requirement is that

$$4c^2 < \log 2. \tag{9.3.26}$$

Therefore with $c = 0.416$, and $\delta = \frac{1}{2}$, formula (9.3.9) implies (9.3.8) and the proof is completed.   **//**

Note that the range of q covers most graphs; that is

$$\sum g_{p,q} \sim g_p \tag{9.3.27}$$

where the sum is over all q which satisfy (9.3.9).

The corresponding theorem for directed graphs is stated next.

Theorem   If for some ε, with $0 < \varepsilon < \frac{1}{2}$,

$$\left|\binom{p}{2} - q\right| = \mathcal{O}(p^{3/2 - \varepsilon}), \tag{9.3.28}$$

then the number $d_{p,q}$ of digraphs with p points and q lines satisfies

$$d_{p,q} \sim \binom{p(p-1)}{q} \bigg/ p!. \tag{9.3.29}$$

The proof is similar to that of (9.3.4) and on carrying out the details, one finds that if

$$\left|\binom{p}{2} - q\right| < 0.832 p^{3/2}, \tag{9.3.30}$$

then (9.3.29) holds.

9.4 CONNECTED GRAPHS AND BLOCKS

Following Riddell [R14], and Ford and Uhlenbeck [FU1] we shall first deal with labeled graphs and then obtain the corresponding results for unlabeled graphs by observing that "most graphs can be considered to be labeled."

Theorem The number G_p of labeled graphs of order p is asymptotic to the number C_p of connected labeled graphs:

$$G_p \sim C_p. \tag{9.4.1}$$

Thus almost all labeled graphs are connected.

Proof On dividing both sides of equation (1.2.1) by G_p, it follows that

$$C_p/G_p = 1 - \frac{1}{p} \sum_{k=1}^{p-1} \binom{p}{k} G_{p-k} C_k/G_p. \tag{9.4.2}$$

Since $C_p \le G_p = 2^{\binom{p}{2}}$ for all p we can write the inequality

$$1 \ge C_p/G_p \ge 1 - F(p), \tag{9.4.3}$$

where

$$F(p) = \frac{1}{p} \sum_{k=1}^{p-1} \binom{p}{k} k 2^{-k(p-k)}. \tag{9.4.4}$$

Therefore to establish (9.4.1) it is sufficient to show that $\lim_{p \to \infty} F(p) = 0$. Since $k/p \le 1$, we have

$$F(p) \le \sum_{k=1}^{p-1} \binom{p}{k} 2^{-k(p-k)}. \tag{9.4.5}$$

Now the summands on the right side of (9.4.5) are end-symmetric with respect to k and hence

$$F(p) \le 2 \sum_{k=1}^{[p/2]} \binom{p}{k} 2^{-k(p-k)}. \tag{9.4.6}$$

But $\binom{p}{k} \le p^k$ and $p - k \ge p/2$ and therefore

$$F(p) \le 2 \sum_{k=1}^{[p/2]} \left(\frac{p}{2^{p/2}} \right)^k. \tag{9.4.7}$$

The sum in (9.4.7) is a geometric series whose common ratio approaches zero as p increases and hence $F(p) = \mathcal{O}(p/2^{p/2})$. Therefore $\lim_{p \to \infty} F(p) = 0$ and the proof is completed. $/\!/$

Wright [W4] has investigated the asymptotic behavior of the coefficients of two generating functions $a(x)$ and $b(x)$ when they satisfy $a(x) = \exp b(x)$. The preceding theorem and other facts can be deduced from his results.

Our next task is to establish the same result as (9.4.1) for unlabeled graphs. But we proceed by first showing that most graphs are *identity graphs*, having only the trivial automorphism group. This fact, known to Riddell [R14], Ford and Uhlenbeck [FU1], Erdös and Rényi [ER1], and others is stated next.

Theorem   The number of graphs of order p is asymptotic to the number of identity graphs of order p, i.e., most graphs of order p can be labeled in $p!$ ways.

Proof   On summing equation (1.1.3) for the number of ways of labeling a graph over all graphs of order p one obtains

$$\sum p!/|\Gamma(G)| = 2^{\binom{p}{2}}. \tag{9.4.8}$$

On dividing both sides of (9.4.8) by $p!$ and applying (9.1.2) we have

$$\sum |\Gamma(G)|^{-1} \sim g_p. \tag{9.4.9}$$

Since the contribution of each graph to the sum in (9.4.9) is less than or equal to 1, most graphs must contribute 1 and the theorem is proved. $/\!/$

As a consequence of this theorem, we have the next far-reaching observation which allows us to obtain some asymptotic results for unlabeled graphs from the corresponding results for labeled graphs and conversely.

Metatheorem   Most labeled graphs have property "P" if and only if most unlabeled graphs have property "P".

From this metatheorem and (9.4.1) follows the companion to (9.4.1) for connected graphs.

Corollary   Almost all graphs are connected.

To establish the next theorem we use some of the results of Section 1.3 on labeled blocks. Recall that B_p is the number of labeled blocks of order p.

Theorem   Almost all labeled graphs are blocks, i.e., $B_p \sim G_p$.

Proof As in Section 1.3, $R(x)$ is the exponential series for rooted, connected, labeled graphs. We let the exponential series $H(x)$, with coefficients H_k, be defined by

$$H(x) = \sum_{k=2}^{\infty} B_k (R(x))^{k-1}/(k-1)! \qquad (9.4.10)$$

From (1.3.8) it follows that $H(x) = R_1(x)/x$. Then on substitution in (1.3.7) and (1.3.4), we have

$$\log \sum_{k=0}^{\infty} C_{k+1} x^k/k! = H(x). \qquad (9.4.11)$$

By the same reasoning used to prove (9.4.1), it can be established that

$$C_{p+1} \sim H_p. \qquad (9.4.12)$$

Since $H(x) = R_1(x)/x$, it also follows that $(p+1)H_p$ is the number of rooted, connected, labeled graphs with $p+1$ points and exactly one block at the root. Hence we can define $F(p)$ by

$$H_p = B_{p+1} + F(p)/(p+1) \qquad (9.4.13)$$

so that $F(p)$ is the number of rooted, connected, labeled graphs with $p+1$ points, exactly one block at the root, and no more than p points in that block. It is easy to see that $F(p)$ has the bound

$$F(p) \le \sum_{k=2}^{p} \binom{p+1}{k} k G_k G_{p+1-k}(p+1-k). \qquad (9.4.14)$$

Now to see that B_p is asymptotic to C_p and hence to G_p it is sufficient to show that

$$\lim_{p \to \infty} F(p)/((p+1)2^{\binom{p+1}{2}}) = 0$$

and this is easily done as in the proof of (9.4.1).

On applying the metatheorem we have the next result.

Corollary Almost all (unlabeled) graphs are blocks.

Although no formal proof is known, we do not hesitate to assert the following statement.

Conjecture For every positive integer n, almost all graphs are n-connected.

From the general methods and results provided in this chapter one can determine the scarcity or preponderance of various classes of graphs. For example, from the formulas (6.2.3) and (6.2.7) for self-complementary graphs and digraphs the following asymptotic determination [P5] can be made using essentially the same method used to prove (9.1.2). Recall that the number of points in a self-complementary graph is congruent to 0 or 1 modulo 4.

Theorem   The numbers $\bar{g}_{4n}$ and $\bar{g}_{4n+1}$ of self-complementary graphs satisfy

$$\bar{g}_{4n} = (2^{2n^2-2n}/n!)(1 + n(n-1)2^{5-4n} + \mathcal{O}(n^3/2^{6n})) \qquad (9.4.15)$$

and

$$\bar{g}_{4n+1} = (2^{2n^2-n}/n!)(1 + n(n-1)2^{4-4n} + \mathcal{O}(n^3/2^{6n})). \qquad (9.4.16)$$

Asymptotic results for self-complementary digraphs are summarized as follows.

Theorem   The asymptotic behavior of the number of self-complementary digraphs can be determined by that of self-complementary graphs, since

$$\bar{d}_{2n} = \bar{g}_{4n} \qquad \text{and} \qquad \bar{d}_{2n+1} \sim 2^n \bar{g}_{4n+1}.$$

We conclude this section by observing the expected result that self-complementary graphs and digraphs are relatively scarce. From equation (9.3.4) we know the asymptotic behavior of the number $g_{4n,r}$ of graphs with $4n$ points and $r = \binom{4n}{2}/2$ lines. With this result and formula (9.4.15), it can be shown that

$$\bar{g}_{4n}/g_{4n,r} \sim 0. \qquad (9.4.17)$$

Corresponding results hold for $\bar{g}_{4n+1}$ and for digraphs. The consequences are stated in the next corollary.

Corollary   Not only are almost all graphs and digraphs not self-complementary, but this holds even for all $(p, p(p-1)/4)$-graphs and correspondingly for digraphs.

9.5   TREES

Pólya [P8] determined asymptotic formulas for the saturated hydrocarbons and other chemical compounds by treating the generating functions as ordinary analytic functions so that the coefficients could be estimated by means of the Cauchy integral formula. Otter [O4] observed that the method could be applied to ordinary trees, and Ford and Uhlenbeck [FU1] found that the method served to approximate the numbers of numerous treelike structures. In this section we shall discuss Otter's results for ordinary trees and hopefully supply enough of the details so that the reader may apply the method to trees of any species.

We require several lemmas, the first of which allows us to treat the generating functions for all trees and rooted trees as analytic functions.

Lemma   The power series $T(x)$ for rooted trees converges in a circle of radius $\eta \geq \frac{1}{4}$.

Proof   The following formula for the number of rooted trees of order $p + 1$ is readily derived from (3.1.9).

$$T_{p+1} = (1/p) \sum_{k=1}^{p} kT_k \sum_{k \leq ks \leq p} T_{p+1-ks}. \tag{9.5.1}$$

But since the coefficients of $T(x)$ are increasing we also have

$$\sum_{k \leq ks \leq p} T_{p+1-ks} \leq (p/k)T_{p+1-k}, \tag{9.5.2}$$

and therefore

$$T_{p+1} \leq \sum_{k=1}^{p} T_k T_{p+1-k}. \tag{9.5.3}$$

Now we define another power series which will bound $T(x)$ above:

$$f(x) = \sum_{k=1}^{\infty} B_k x^k \tag{9.5.4}$$

where $B_1 = 1$, and for $p \geq 1$

$$B_{p+1} = \sum_{k=1}^{p} B_k B_{p+1-k}. \tag{9.5.5}$$

It follows by induction from (9.5.3) that $T_p \leq B_p$ for all p. But $y = f(x)$ satisfies

$$y^2 - y + x = 0. \tag{9.5.6}$$

On solving this equation for y as a function of x, one obtains

$$y = \tfrac{1}{2}(1 \pm (1 - 4x)^{1/2}). \tag{9.5.7}$$

and since $f(0) = 0$, we have

$$f(x) = \tfrac{1}{2}(1 - (1 - 4x)^{1/2}). \tag{9.5.8}$$

When this is expanded, using the binomial theorem, into a power series in x, the coefficients are precisely those in equation (3.3.23) which counts planted plane trees, as well as all other configurations enumerated by Catalan numbers. On developing the Taylor series for $f(x)$ at the origin we can conclude that for each p

$$B_p = \frac{1}{2}\binom{\frac{1}{2}}{p}(-1)^{p-1}4^p, \tag{9.5.9}$$

thus establishing a bound for T_p. Furthermore, it follows from (9.5.8) that the radius of convergence of the series $\sum_{k=1}^{\infty} B_k x^k$ is $\frac{1}{4}$ and hence that of $T(x)$ is at least $\frac{1}{4}$.

By virtue of formula (3.2.4) which expresses the series $t(x)$ for trees in terms of $T(x)$, an immediate consequence of this lemma is that $T(x)$ and $t(x)$ have the same radius of convergence! In addition, $T(x)$ and $t(x)$ bound the corresponding series for trees of any species, and hence the series for the latter also converge. These facts were first found by Pólya [P8].

At this point we observe that since all the coefficients of the generating function $T(x)$ are positive, η is a singularity of $T(x)$ and hence of $t(x)$. On the other hand, $T(x)$ converges with $x = \eta$. To establish this fact we need the next statement.

Lemma   The limit of $T(x)$ as $x \to n-$ exists and is equal to $\sum_{k=1}^{\infty} T_k \eta^k$.

Proof   Since $T(x)$ satisfies the functional equation (3.1.4), we have for all x in $(0, \eta)$

$$\log(T(x)/x) = T(x) + \sum_{k=2}^{\infty} T(x^k)/k. \tag{9.5.10}$$

From this it follows that

$$\frac{T(x)/x}{\log(T(x)/x)} \leq \frac{1}{x}, \tag{9.5.11}$$

and hence $T(x)$ is bounded on the interval $(0, \eta)$. Since $T(x)$ is monotone, the left-hand limit at η exists and we let

$$b_0 = \lim_{x \to \eta-} T(x). \tag{9.5.12}$$

It now follows quickly that $b_0 = T(\eta)$.   //

The value of b_0 is determined by the next lemma.

Lemma The series $T(x)$ for rooted trees has the property that

$$T(\eta) = 1. \tag{9.5.13}$$

Proof First we define the complex valued function $F(x, y)$ for complex x and y by

$$F(x, y) = x \exp\left\{ y + \sum_{k=2}^{\infty} T(x^k)/k \right\} - y \tag{9.5.14}$$

and consider the equation

$$F(x, y) = 0. \tag{9.5.15}$$

From (3.1.4) we can show that $y = T(x)$ is the unique analytic solution of (9.5.15), and we know it has a singularity at $x = \eta$. The preceding lemma implies that $F(\eta, b_0) = 0$ and furthermore it is easy to see that $F(x, y)$ is analytic in each variable separately in neighborhoods of η and b_0.

On differentiating (9.5.14) with respect to y, we find

$$\partial F/\partial y = F(x, y) + y - 1. \tag{9.5.16}$$

Since $F(\eta, b_0) = 0$, we know that this partial derivative at (η, b_0) is given by

$$\frac{\delta F}{\delta y}(\eta, b_0) = b_0 - 1. \tag{9.5.17}$$

Furthermore, this partial derivative must be zero at (η, b_0), i.e. $b_0 = 1$. Otherwise, by the implicit function theorem, there is a unique solution $y = f(x)$ of (9.5.15) which is analytic in a neighborhood of η, in particular at η itself. But such a solution would have to be $y = T(x)$ and we know that the latter is *not* analytic at $x = \eta$, proving (9.5.13).   //

Note that it follows immediately from (9.5.16) that the second partial derivative of $F(x, y)$ with respect to y is not zero at $(\eta, 1)$.

For the sake of brevity, we omit the proof of the next theorem which is a combination of the implicit function theorem and observations of Pólya, Otter, Ford, and Uhlenbeck.

Theorem Let $F(x, y)$ be analytic in each variable separately in some neighborhood of (x_0, y_0) and suppose that the following conditions are satisfied:

  i. $F(x_0, y_0) = 0$;

 ii. $y = f(x)$ is analytic in $|x| < |x_0|$ and x_0 is the unique singularity on the circle of convergence;

iii. if $f(x) = \sum_{n=0}^{\infty} f_n x^n$ is the expansion of f at the origin, then
$y_0 = \sum_{n=0}^{\infty} f_n x_0^n$;

iv. $F(x, f(x)) = 0$ for $|x| < x_0$;

v. $\dfrac{\partial F}{\partial y}(x_0, y_0) = 0$;

vi. $\dfrac{\partial^2 F}{\partial y^2}(x_0, y_0) \neq 0.$

Then $f(x)$ may be expanded about x_0:

$$f(x) = f(x_0) + \sum_{k=1}^{\infty} a_k(x_0 - x)^{k/2} \tag{9.5.18}$$

and if $a_1 \neq 0$,

$$f_n \sim \frac{-a_1}{2\sqrt{\pi}} x_0^{-n+1/2} n^{-3/2} \tag{9.5.19}$$

and if $a_1 = 0$ but $a_3 \neq 0$

$$f_n \sim \frac{3a_3}{4\sqrt{\pi}} x_0^{-n+3/2} n^{-5/2}. \tag{9.5.20}$$

To apply this theorem, note that the function defined by (9.5.14) satisfies all the hypotheses with $(x_0, y_0) = (\eta, 1)$ and $f(x) = T(x)$. Hence $T(x)$ may be expanded as in (9.5.18) and if $a_1 \neq 0$, the coefficients behave as in (9.5.19). It remains to be shown that if

$$T(x) = 1 - b_1(\eta - x)^{1/2} + b_2(\eta - x) + b_3(\eta - x)^{3/2} + \cdots \tag{9.5.21}$$

then $b_1 \neq 0$ and $b_3 \neq 0$, and, of course we require approximations to b_1 and η.

On differentiating (9.5.21) we have

$$T'(x) = \tfrac{1}{2}b_1(\eta - x)^{-1/2} - b_2 + \cdots \tag{9.5.22}$$

where the terms omitted contain $(\eta - x)^{1/2}$ to the first power at least. On multiplying both sides of (9.5.22) by $1 - T(x)$ as obtained from (9.5.21) we have

$$T'(x)(1 - T(x)) = \tfrac{1}{2}b_1^2 + \cdots \tag{9.5.23}$$

where, again, the terms omitted contain $(\eta - x)^{1/2}$ to at least the first power. Hence

$$\lim_{x \to \eta-} T'(x)(1 - T(x)) = \tfrac{1}{2}b_1^2. \tag{9.5.24}$$

On differentiating (3.1.4), however, one obtains

$$T'(x) = T(x)/x + T(x) \sum_{k=1}^{\infty} T'(x^k)x^{k-1}, \qquad (9.5.25)$$

and therefore

$$T'(x)(1 - T(x)) = T(x)/x + T(x) \sum_{k=2}^{\infty} T'(x^k)x^{k-1}. \qquad (9.5.26)$$

Thus the limit in (9.5.24) can also be obtained from (9.5.26) and hence

$$\tfrac{1}{2}b_1^2 = \frac{1}{\eta} + \sum_{k=2}^{\infty} T'(\eta^k)\eta^{k-1}. \qquad (9.5.27)$$

Using (3.1.4) and (9.5.13), Otter estimated $\eta = 0.3383219$. Then from an equation similar to (9.5.27) he estimated b_1 and found that

$$\frac{b_1\eta^{1/2}}{2\sqrt{\pi}} = 0.4399237\cdots \qquad (9.5.28)$$

with $b_1 = 2.681127$. Therefore, making use also of (9.5.20), the behavior of the number of rooted trees can be stated as follows.

Theorem   The number T_p of rooted trees of order p satisfies

$$T_p = 0.4399237\frac{\eta^{-p}}{p^{3/2}} + \mathcal{O}\left(\frac{\eta^{-p}}{p^{5/2}}\right). \qquad (9.5.29)$$

Now we consider the series $t(x)$ for (unrooted) trees. From (3.2.4) and (9.5.21) it follows that we may write

$$t(x) = a_0 - a_1(\eta - x)^{1/2} + a_2(\eta - x) + a_3(\eta - x)^{3/2} + \cdots, \quad (9.5.30)$$

hence

$$t'(x) = \tfrac{1}{2}a_1(\eta - x)^{-1/2} - a_2 - \tfrac{3}{2}a_3(\eta - x)^{1/2} + \cdots. \qquad (9.5.31)$$

On differentiating (3.2.4), however, we find

$$t'(x) = T'(x)(1 - T(x)) + T'(x^2)x \qquad (9.5.32)$$

and from (9.5.24) it follows that $\lim_{x\to\eta_-} t'(x)$ exists. Hence $a_1 = 0$. Next we express the second derivative $t''(x)$ in two ways. From (9.5.31) with $a_1 = 0$ we have

$$t''(x) = \tfrac{3}{4}a_3(\eta - x)^{-1/2} + \cdots. \qquad (9.5.33)$$

TABLE 9.5.1

APPROXIMATIONS FOR THE NUMBER OF TREES

p	T_p	$\tilde{T}_p$	t_p	$\tilde{t}_p$
1	1	1	1	2
2	1	1	1	1
3	2	2	1	1
4	4	4	2	2
5	9	9	3	2
10	719	708	106	86
15	87 811	86 965	7 741	7 050
18	1 721 159	1 708 440	123 867	114 875

On substituting the right side of (9.5.26) into (9.5.32) and differentiating, we find that

$$t''(x) = T'(x)\left\{1/x + \sum_{k=2}^{\infty} T'(x^k)x^{k-1}\right\} + \cdots, \qquad (9.5.34)$$

where the terms omitted are bounded near η. Then from (9.5.22) and (9.5.27) we have

$$t''(x) = \tfrac{1}{4}b_1^3(\eta - x)^{-1/2} + \cdots, \qquad (9.5.35)$$

and therefore $a_3 = b_1^3/3$. Then the behavior of the number of trees can be deduced from (9.5.20) on finding $(b_1^3/4\sqrt{\pi})\eta^{3/2} = 0.5349485\ldots.$

Theorem The number t_p of trees of order p satisfies

$$t_p = 0.5349485\frac{\eta^{-p}}{p^{5/2}} + \mathcal{O}\left(\frac{\eta^{-p}}{p^{7/2}}\right). \qquad (9.5.36)$$

We conclude with Otter's table (9.5.1) which compares t_p and T_p with the values given by (9.5.36) and (9.5.29) without the error terms denoted by $\tilde{t}_p$ and $\tilde{T}_p$.

EXERCISES

9.1 (a) Relations, asymptotically.   (Oberschelp [O1])
     (b) Tournaments, asymptotically.   (Moon [M2])

9.2 Connected (p, q) graphs (labeled and unlabeled), asymptotically.
                                        (Erdös and Renyi [ER1]; Wright [W6])

9.3   Investigate the relation

$$1 + \sum_{n=1}^{\infty} \frac{F_n}{n!} x^n = \exp\left(\sum_{n=1}^{\infty} \frac{f_n}{n!} x^n\right)$$

to determine how rapidly f_n must increase to insure that $\lim_{n \to \infty} F_n/f_n = 1$.
(*Hint:*   Read the Wright papers.)   (Wright [W4])

9.4   For any positive k, the number g_p of graphs satisfies

$$g_p = \frac{2^{\binom{p}{2}}}{p!}\left(1 + \sum_{i=1}^{k} \varphi_i(p)2^{-kp} + \mathcal{O}\left(\frac{p^{2k}}{2^{kp}}\right)\right),$$

where $\varphi_i(p) = p(p-1)\cdots(p-i)\Psi_i(p)$ and $\Psi_i(p)$ is a polynomial of degree
$i-1$ in p.   (Wright [W5])

9.5   The number of labeled graphs of order p with no triangles is asymptotic
to $2^{(p^2/4)}(1 + o(1))$.   (D. Kleitman)

9.6   The   number $\bar{t}_{2n}$ of   self-complementary   tournaments   equals
$(2^{n^2-n}/n!)(1 + o(1))$.

Self-complementary tournaments of odd order, asymptotically.

9.7   Almost all:
(a)   tournaments are strong;   (Moon and Moser [MM1])
(b)   digraphs are strong;   (Moon and Moser [MM2])
(c)   digraphs are hamiltonian;   (Moon [M5])
(d)   graphs are hamiltonian.   (Moon [M5])

9.8   Connected labeled functional digraphs of order p are asymptotic to
$(\pi/2)^{1/2}p^{p-1/2}$ (Rényi [R13]; Katz [K2])

9.9   Pure 2-dimensional simplicial complexes of order p are asymptotic to
$2^{\binom{p}{3}}/p!$.

9.10   The probability that a random point of the labeled trees of order p
is an endpoint approaches $1/e$ as p increases.   (Rényi [R12])

9.11   Almost all trees have nontrivial automorphisms.
                                                              (Ford and Uhlenbeck [FU1])

9.12   A sufficient condition for (9.3.4) is $\min(q, \binom{p}{2} - q) \geq 3p \log p$.
                                                                      (Wright [W7])

9.13   A necessary and sufficient condition for (9.3.4) is

$$\lim_{p \to \infty} \left(\frac{2\min(q, \binom{p}{2} - q)}{p} - \log p\right) = \infty.$$

                                                                      (Wright [W7])

Chapter 10 | UNSOLVED PROBLEMS

Although various sophisticated, recondite, and specialized terminology may confuse the situation, the fact is that very many pattern and configuration problems become graphical in nature when properly reformulated. Furthermore, the conceptual difficulty of the problem is more easily identified when recast in terms of graphs or variations on graphs. We present a wide range of graphical enumeration problems, containing adequate material to occupy research scholars for generations.

There exist several earlier lists of exactly twenty seven unsolved problems in graphical enumeration. That restriction is now abandoned and myriads of open questions are exposed. In this and the remaining sections, we list the problem areas as P1.1, P1.2, ... Each such area in turn may contain several individual problems which will be indicated. In some cases, data in the form of the first few terms of the generating function will be included.

We include no explicit problems on counting trees of various types, for the existing methods appear always adequate. For counting labeled trees, see Moon [M4] and Chapter 1, and for unlabeled trees, see Chapter 3.

10.1 LABELED GRAPHS

The succeeding sections include counting questions open for both labeled and unlabeled graphs. Usually the labeled case is more manageable than the unlabeled case because there is less symmetry involved. However there are two notable exceptions for which unlabeled configurations have been counted but not the labeled ones:

P1.1　*Labeled self-converse digraphs*
The unlabeled self-converse digraphs were counted in Chapter 6 using the Power Group Enumeration Theorem.

P1.2　*Labeled self-complementary graphs*
The number of self-complementary graphs and digraphs was determined by Read [R6], see Chapter 6.

The labeled cases for both of these problems are untouched.

10.2 DIGRAPHS

There are many unsolved problems involving digraphs which are better stated later along with the corresponding problems for graphs in the succeeding sections. Nevertheless there are some which merit separate mention here because they involve structural properties exclusive to digraphs.

P2.1　*Strong digraphs*
Recall from Chapter 5 that a digraph is *strong* if each pair of points are mutually reachable by directed paths. Our good friend R. W. Robinson succeeded in enumerating both labeled and unlabeled strong digraphs several years ago, but has not yet found the time to write up these interesting, important, and difficult results. It is to be hoped that he will do so within the present decade. His methods involve the condensation D^* of an arbitrary digraph D in which the points of D^* are the strong components of D, together with his techniques of cycle index sums developed in Chapter 8.

Figure 10.2.1

The strong digraphs of order 3.

All the digraphs of order 4 are listed in [H1, pp. 227–330]. There are 83 strong digraphs of order 4 as indicated in [HP6]. Hence their counting series begins:

$$x + x^2 + 5x^2 + 83x^4 + \cdots.$$

P2.2   *Unilateral digraphs*

A digraph is *unilaterally connected* or *unilateral* if for any two points, at least one is reachable from the other. Although its unilateral components do not partition a digraph, the method of Robinson mentioned above for counting strong digraphs can still be specialized to unilateral digraphs, whose counting series as calculated by R. C. Read begins

$$x + 2x^2 + 11x^3 + 172x^4 + 8603x^5 + \cdots.$$

But by a theorem in [HNC1, p. 66], a digraph is unilateral if and only if its condensation contains a unique spanning path. For this reason, a suitable modification of the counting of strong digraphs might serve to count unilateral digraphs, and this is in fact Robinson's approach which he has not as yet published or even written.

P2.3   *Digraphs with a source*

A point in a digraph is a *source* if all other points can be reached from it. The directional dual is a *sink*. Of course there are the same number of digraphs with a source as with a sink, as these are converse collections. The series for these starts

$$x + 2x^2 + 12x^3 + 184x^4 + \cdots.$$

In a strong digraph, every point is both a source and a sink. The counting of digraphs with both a source and a sink is also open. Robinson also claims that he can count these kinds of digraphs.

P2.4   *Transitive digraphs*

A digraph is *transitive* if the presence of arcs uv and vw implies that of arc uw. It is very easy to see that transitive digraphs of order p correspond precisely to finite topologies on a set of p elements. These have only been enumerated in the labeled case [EHL1], using Stirling numbers of the second kind.

The first four terms in the series for unlabeled transitive digraphs are

$$x + 3x^2 + 9x^3 + 32x^4 + \cdots.$$

Even the next term is unknown. For the labeled case, the first nine terms have been computed, and each coefficient takes exponentially more computer time; see Appendix I.

P2.5 *Digraphs both self-complementary and self-converse*
The only digraphs of order 3 that are both self-complementary and self-converse are the cyclic and transitive triples. The counting series starts

$$x + x^2 + 2x^3 + 4x^4 + \cdots.$$

This is an interesting new type of problem which seems to call for an appropriate generalization of Burnside's Lemma.

P2.6 *Eulerian digraphs*
Eulerian graphs have been counted also by Robinson [R18], but the techniques are not adaptable to the corresponding problem for digraphs. The counting series for eulerian digraphs begins

$$x + x^2 + 3x^3 + 12x^4 + 68x^5 + \cdots.$$

A. Kotzig raised the special case of this question for eulerian tournaments, whose series begins

$$x + x^3 + x^5 + 3x^7 + \cdots.$$

10.3 GRAPHS WITH GIVEN STRUCTURAL PROPERTIES

In this book we have shown how to count many graphs with specified structural properties. Typical examples involving cycles include trees, unicyclic graphs, and functional digraphs. Concerning connectivity, we have counted connected graphs, blocks and block-graphs. We shall consider in this section eight categories of graphs:

a. hamiltonian
b. eulerian (p, q) graphs
c. graphs with local subgraphs
d. identity graphs
e. symmetric graphs
f. graphs with a square root
g. line and total graphs
h. clique and interval graphs

Each of these contains in turn several individual counting problems, so that there are many questions in this section. Furthermore, whenever a new structural property is introduced into the graphical literature, one can take this as a challenge to find how many such graphs exist.

P3.1 *Hamiltonian graphs*
A graph or digraph is *hamiltonian* if it contains a spanning cycle. These have not been counted for labeled or unlabeled graphs or digraphs. The series for unlabeled graphs begins

$$x^3 + 3x^4 + 8x^5 + 48x^6 + \cdots,$$

while that for unlabeled digraphs starts

$$x^2 + 4x^3 + 60x^4 + \cdots.$$

This is the most publicized special case of graphs containing a specified subgraph, namely a spanning cycle. One can also stipulate graphs with other kinds of subgraphs such as a 1-factor, a 1-basis, cycles of given length, or complete graphs of given order. We will elaborate on this theme in Section 10.6.

P3.2 *Eulerian graphs*

Robinson [R18] counted eulerian graphs of order p without regard for the number of lines, see Section 4.7. It is most difficult to count eulerian (p, q) graphs, where the number of lines is also an enumeration parameter. For eulerian graphs of order 6, the counting polynomial, where the exponent gives the number of lines, is

$$x^6 + x^7 + 2x^8 + x^9 + x^{10} + x^{11} + x^{12}.$$

Another interesting parameter for eulerian graphs would involve the minimum number of cycles whose union is the entire graph. This stems from the theorem, see [H1, p. 64], that a connected graph is eulerian if and only if its set of lines can be partitioned into cycles.

P3.3 *Local subgraphs*

Given a graph H, the problem is to find the number of graphs of order p such that each point lies in a subgraph isomorphic to H. For example, if H is a triangle, then the series begins

$$x^3 + 2x^4 + 7x^5 + 37x^6 + \cdots.$$

We also ask for the number of graphs in which each line lies in a triangle. Of course, similar questions suggest themselves for digraphs.

P3.4 *Identity graphs*

There are no nontrivial graphs of order less than 6 which have the identity group. For $p = 6$, there are eight identity graphs. We have seen in Section 9.4 that asymptotically most graphs have the identity group, but there is no exact formula for order p. Considering the same problem for digraphs, the series begins

$$x + x^2 + 7x^3 + 137x^4 + \cdots.$$

Identity trees have been counted [HP14], and also identity unicyclic graphs and identity functional digraphs by Stockmeyer [S6]. In principle,

Stockmeyer obtained a formula for the number of graphs with given automorphism group. But its use entails the knowledge of the entire lattice of subgroups of the symmetric group S_p. Thus while this theoretically includes the enumeration of the number of identity graphs and digraphs, it cannot be properly regarded as a solution to this problem (see Table A10 in Appendix II).

P3.5 *Symmetric graphs*

In a point-symmetric graph, the automorphism group is transitive on the points. A *line-symmetric* graph is then defined as expected. A *symmetric* graph is both point-symmetric and line-symmetric.

Turner [T1] counted point-symmetric graphs with a prime number p of points only. Chao [C3] proved that there exists a symmetric graph of prime order p, regular of degree n, if and only if n is even and divides $p - 1$; furthermore such a graph is unique up to isomorphism. No other results are known about the number of symmetric graphs.

P3.6 *Graphs with a square root*

The *square G^2* of a graph G has the same points as G, with u and v adjacent in G^2 whenever their distance in G is either 1 or 2. Other powers G^3, G^4, ... are defined similarly. Graphs which have a square root have been characterized, see [H1, p. 24], and similar results were found for digraphs by Geller [G1]. The counting of graphs and digraphs which have an nth root may not be an impossible problem.

P3.7 *Line graphs and total graphs*

The *line graph $L(G)$* has the lines of graph G as its points with adjacency of lines as in G. A *line graph* is the line graph of some graph. This concept was introduced by Whitney [W2] who showed that a line graph H is the line graph of only one graph unless $H = K_3$. Thus the number of connected line graphs of order $p > 3$ is exactly the number of connected graphs with p lines. From the table in Cadogan [C1] giving the number of connected graphs, the generating function for connected line graphs with a given number of points begins

$$x + x^2 + 2x^3 + 5x^4 + 12x^5 + 30x^6 + 79x^7 + 227x^8 + \cdots.$$

The problem here is to find a more direct method, as well as to count line graphs with a given number of points *and* lines. Such an approach would probably use one of the structure theorems for line graphs given in [H1, Chapter 8].

Connected digraphs with q arcs are not the same in number as connected line digraphs of order q, as shown in [HN3]. Hence we do not even have this circuitous approach to the counting of line digraphs.

The *total graph* $T(G)$ has the points and lines of G as its point set, with adjacency defined more or less as expected; see [H1, p. 82]. It is known that the total graph $T(G)$ is the square of the subdivision graph $S(G)$ obtained by inserting a new point of degree 2 into each line of G. But this does not seem to facilitate the search for a formula.

P3.8   *Clique graphs and interval graphs*

A *clique* of a graph G is a maximal complete subgraph. The *clique graph* of G has the cliques of G as its points, with adjacency determined by non-empty intersection of two cliques. An *interval graph* has intervals on the real line as its points, with adjacency again determined by intersection. In a *rigid circuit graph*, there are no induced cycles of length greater than three. These classes of graphs are known to be related to each other by their characterization theorems. For example, clique graphs were characterized structurally by Roberts and Spencer [RS1], as reported in [H1, p. 20].

10.4 GRAPHS WITH GIVEN PARAMETER

It is much easier to propose enumeration problems than to solve them. For example, each time a new parameter is discovered our viewpoint immediately formulates an associated counting question. Counting problems for graphs with given parameters partition themselves naturally into sets of parameters which are closely related. Our ten categories are:

a. radius and diameter
b. girth and circumference
c. minimum and maximum
    degrees
d. connectivity
e. independence and covering
    numbers

f. clique numbers
g. intersection number
h. arboricity
i. genus and thickness
j. chromatic numbers

For each of these categories, we define the related invariants and describe the partial progress which has been attained.

P4.1   *Radius and diameter*

The *eccentricity* $e(v)$ of a point v of a graph G is the maximum distance between v and any other point. The *radius* $r(G)$ is the minimum eccentricity in G and the *diameter* $d(G)$ is the maximum.

Thus a graph G has radius 1 if and only if it has a point v_0 adjacent to all other points. It is easy to tell the number of graphs of order p with radius 1 because this number is precisely the total number of graphs of order $p - 1$. Even for graphs with radius 2, there are no immediate solutions.

For trees, the diameter is approximately double the radius. Trees with prescribed diameter d were counted in [HP14]. For graphs there is just the trivial observation that the only p-point graph with $d = 1$ is K_p.

For digraphs, there are corresponding problems. The inradius and out-radius of a digraph D were defined in [HNC1, p. 162]. These invariants as well as the diameter always exist when D is strong.

P4.2 *Girth and circumference*

The *girth* of a graph is the minimum length of a cycle in it; the *circumference* the maximum. Let c_n be the number of cycles of length n in a graph G. How many graphs are there with such a prescribed sequence $(c_3, c_4, c_5, \ldots, c_p)$? The answer to this one general question would count the following structures with only a little more trouble:

a. graphs with given girth, and *a fortiori*,
b. graphs with given circumference,
c. graphs containing a triangle,
d. graphs containing a quadrilateral,
e. hamiltonian graphs,
f. graphs with a given number of cycles.

The counting of unicyclic graphs in Section 3.4 is a very minor step in this direction.

P4.3 *Minimum and maximum degrees*

The minimum degree δ and the maximum degree Δ are natural parameters to consider for counting problems. Although there are theoretical formulations for counting graphs with a given partition, Parthasarathy [P7], and digraphs with given partitions [HP3], these do not bear directly on δ and Δ.

Consider the graphs with $\delta \geq n > 0$. When $n = 1$, these are all the graphs with no isolated points, and are readily reckoned. The case $n = 2$ comprises all graphs with no endpoints as well as no isolates. These were counted by Robinson [R19] using the method of cycle index sums developed for counting blocks; see Section 8.6. For $n = 3$, these are the homeomorphically irreducible graphs with no end points.

P4.4 *Connectivity*

The *connectivity* κ (*line-connectivity* λ) of a graph G is the minimum number of points (lines) whose removal from G results in a graph which is either disconnected or trivial. Then G is *n-connected* if $\kappa \geq n$. Thus the number of graphs with connectivity n is the number of n-connected graphs minus the number of $(n + 1)$-connected graphs.

Both 0-connected and 1-connected graphs have been counted since these are the disconnected and connected graphs respectively. The 2-connected graphs with $p \geq 3$ are the same as blocks and these were counted by Robinson, see Chapter 8. The enumeration of n-connected graphs for $n \geq 3$ evidently requires more powerful methods than now exist.

One can also ask for the number of graphs with given κ and λ, and as a special case for the number with $\kappa = \lambda$.

P4.5   *Independence and covering numbers*

A set of points (lines) is *independent* if no two are adjacent. The *point-independence number* is the maximum number β_0 of independent points of G, and the *line-independence number* β_1 is defined similarly. A point v and a line x *cover each other* if v is on x. The *point-covering number* α_0 of G is the minimum number of points which cover all the lines, and the *line-covering number* α_1 switches points and lines.

Since it is a classic equation in graph theory that $\alpha_0 + \beta_0 = p = \alpha_1 + \beta_1$, we need only mention the problems of counting graphs with given point-independence number β_0 and with given β_1. Intuitively these seem easier than counting graphs with given covering numbers, but by the above equation there is no difference.

P4.6   *Clique numbers*

There are several invariants associated with cliques. One of these is the greatest clique order of G which we have just encountered in the form $\beta_0(\bar{G})$, the maximum number of independent points in the complement of $\bar{G}$. Some other clique numbers are:

a.   the number of cliques
b.   the minimum order of a clique
c.   the minimum number of cliques which cover all the points of G
d.   similarly for covering the lines of G
e.   the maximum number of point-disjoint cliques.

This last invariant is of course the point-independence number of the clique graph of G, while the first is the number of points in this clique graph. In general, any invariant of the clique graph of G becomes in this way an invariant of G itself. None of these problems seem promising.

P4.7 *Intersection number*

The *intersection number* $\omega(G)$ of a given graph G is the minimum number of elements in a set S such that there is a family $S_1, S_2, \ldots, S_p$ of distinct, nonempty subsets of S whose union is S and v_i and v_j are adjacent in G if and only if $S_i \cap S_j \neq \varnothing$. Another variation on this invariant is ω_0 which differs from ω only in that the sets S_i need not be distinct, so that for example $\omega_0(K_p) = 1$.

There does not seem to exist any method for counting graphs with given intersection number or any of its possible variations.

P4.8 *Arboricity*

The *arboricity* of a graph G is the minimum number of line-disjoint acyclic subgraphs whose union is G. The maximum number of line-disjoint nonacyclic subgraphs whose union is G is called the *anarboricity* of G. As noted in [H13], these constitute a pair of covering (arboricity) and packing (anarboricity) invariants. Another such pair is given by the covering path number, which is the smallest number of paths whose union is G, and its packing counterpart. Such invariants appear hopeless for use as enumeration parameters, as does also the number of spanning trees in a graph, called its *complexity* by Brooks, Smith, Stone, and Tutte [BSST1].

P4.9 Genus, thickness, coarseness, crossing number

These and other topological invariants may well be the most intractable of all as far as enumeration is concerned. The *genus* γ of G is the minimum genus of an orientable surface on which G can be embedded with no pair of edges intersecting. The *thickness* θ is the smallest number of planar subgraphs whose union is G. The *coarseness* ξ is the greatest number of line-disjoint nonplanar subgraphs whose union is G. And the *crossing number* v is the smallest number of pairs of edges which intersect when G is drawn in the plane. When G is planar, $\gamma = v = 0$, $\theta = 1$, and ξ is not defined. There are many other related topological invariants of a graph, each at least as hopeless for counting as the above four.

P4.10 *Chromatic number*

The *chromatic number* χ of a graph G is the minimum number of colors needed for its points so that no two adjacent points have the same color. The *line-chromatic number* χ' (*total-chromatic number* χ'') of G is the chromatic number of the line graph (total graph) of G. (Some authors call χ' the chromatic index of a graph.)

A graph G is *n-chromatic* if $\chi = n$, and G is *n-colorable* if $\chi \leq n$. The n-colored graphs were counted by Robinson [R17], see Section 4.5, and the 2-colorable graphs in [HP15]. Even the 3-colorable graphs appear impossible to enumerate at present.

10.5   SUBGRAPHS OF A GIVEN GRAPH

Most of the problems in this section ask for the number of dissimilar subgraphs of a given graph G that are isomorphic to a certain graph H. Thus the group of G determines whether or not two occurrences of subgraph H are regarded as equivalent. Analogous questions can also be posed for digraphs.

P5.1   *Hamiltonian cycles*
    The number of spanning cycles in a given graph or digraph can theoretically be expressed in the labeled case using the method of Cartwright and Gleason [CG1] in terms of the adjacency matrix. But the calculation of such numbers is forbidding, and becomes even more involved when the number of similarity classes is desired. Only for special graphs such as K_p can the answer be written from first combinatorial principles.
    The most interesting special case of this problem arises when the given graph is the n-cube Q_n because of applications to coding theory. It is easily observed that there is only one similarity class of spanning cycles in Q_2 and Q_3. Gilbert [G2] has shown that the series for these begins

$$x^2 + x^3 + 9x^4 + \cdots,$$

but none of the coefficients are known for $n > 4$. The number of labeled hamiltonian cycles in Q_n has also been found only for $n \leq 4$. An undetermined amount of computer time is required to calculate just the next coefficient.

P5.2   *Cycles of a given length*
    The problem asks for the number of dissimilar cycles of length k in a graph of order p, generalizing P5.1. It is easily solved in special cases. For example, there is just one similarity class of cycles of length $2k$ in the complete bipartite graph $K_{m,n}$ when $m, n \geq k$. The answer to the labeled version of the same question is $\binom{m}{k}\binom{n}{k}k!(k-1)!/2$. A solution of the labeled problem for cycles of length 3, 4, and 5 in a given graph or digraph in terms of the adjacency matrix was found in [HM1].

P5.3   *Complete graphs*
    As mentioned above, the number of triangles K_3 in a given labeled graph can be calculated. The problem of determining the number of occurrences of K_n, $n > 3$, in a given labeled graph is open. For the unlabeled case, the knowledge of not only the total number of triangles, but also their behavior with respect to the group of the given graph is required to obtain the number of dissimilar triangles.

P5.4 *Spanning trees*

The number of labeled spanning trees of a given labeled graph can be calculated using the Matrix-Tree Theorem. But there is no corresponding method for determining the number of different (nonisomorphic) spanning trees of a given graph. The difficulty of the problem is indicated by the example in which the given graph is the complete graph of order p. Then the number of different spanning trees is the number of trees of order p, a problem first solved by Cayley. The counting of the number of dissimilar spanning trees in a given graph is also open.

P5.5 *Factors*

Let G be a labeled graph having a 1-factor, that is, a spanning subgraph of independent lines; see [H1, p. 86]. The number of different 1-factors of G is not known except in very special cases. For example, K_{2n} has $(2n)!/2^n n!$ 1-factors and $K_{n,n}$ has just $n!$ of them.

A *factorization* of G, if any, is a partition of the lines of G into 1-factors. It is known that K_{2n} has a 1-factorization but the number of these has not been settled except for small n. Recently W. Wallis established that K_8 has exactly six 1-factorizations. By an older theorem of König, see [H1, p. 85], every regular bigraph such as $K_{n,n}$ has a 1-factorization, so that the same question can be asked for these.

P5.6 *Eulerian trails in a given eulerian graph*

There is an explicit formula for the number of eulerian trails in a given digraph, see (1.8.2). For graphs, however, no progress has been made. One possible approach would be to consider all orientations of a given eulerian graph G which result in eulerian digraphs $D_1, D_2, \ldots$. If e_i is the number of eulerian trails in D_i, then $\sum e_i$ is the total number of oriented eulerian trails G. But this is easier said than done. For example, the special case $G = K_{2n+1}$ requires the availability of the adjacency matrices of all eulerian tournaments of order $2n + 1$.

10.6 SUPERGRAPHS OF A GIVEN GRAPH

The problems in this section ask for the number of graphs of order p that are supergraphs of a given graph H. Extremal graph theory may be useful in counting such graphs. For example, with $H = K_3$, Turán's theorem shows that if G has at least $[p^2/4]$ lines then G must contain K_3. Therefore the counting problem need only be handled for graphs with less than $[p^2/4]$ lines. On the other hand, a solution to such a counting problem may solve the corresponding extremal problem. Versions of all of these problems also exist for digraphs.

P6.1   Cycles

Counting supergraphs of a triangle is the same as counting graphs of girth 3. But supergraphs of the cycle C_n of order $n \geq 4$ do not correspond to graphs of girth n because smaller cycles are not excluded. The series for supergraphs of C_4 begins

$$3x^4 + 16x^5 + 111x^6 + \cdots.$$

P6.2   Complete graphs

The problem of counting the number of supergraphs of K_n has already arisen in different guises for $n = 3$. Therefore our interest here focuses on $n \geq 4$. Turán's general theorem solves the associated extremal problem, see [H1, p. 18].

P6.3   Complete bipartite graphs

The problem is to count supergraphs of $K_{m,n}$. Since $K_{2,2}$ is a cycle of order 4, we have already encountered a difficult special case. Observe that the supergraphs of $K_{1,n}$ are those graphs whose maximum degree is at least n; compare problem P4.3.

P6.4   Paths

Counting supergraphs of the path P_n of order n is easy for small n. For example if $n = 4$, the only connected graphs of order ≥ 4 which are *not* supergraphs of P_n are the stars $K_{1,m}$. A solution for each n would involve knowledge of those connected graphs of order $\geq n$ whose spanning trees all have diameter less than n. This problem is related to P5.4.

10.7   GRAPHS AND COLORING

There exists a *bona fide* method for settling the 4CC (Four Color Conjecture) which depends on the solution of certain graphical enumeration problems:

4CC: Every planar graph is 4-colorable.
EE4CC: The number of planar graphs equals the number of 4-colorable planar graphs. (EE4CC stands for the Enumeration Equivalent of the 4 Color Conjecture.)

The counting series for these two classes of graphs are known to have the same first 39 coefficients; see [OS1]. There is some latitude with regard to two different degrees of freedom:

1. The parameter can vary. One can use either p points, or p points and q lines, or q lines, or diameter d, or any other parameter for which there is some hope.

2. The type of graph can vary provided only that both the planar graphs and the 4-colorable planar graphs share the same properties, which may include:

a. planar graphs (as in EE4CC)
b. rooted planar graphs
c. line-rooted planar graphs
d. planar graphs rooted at a triangle
e. labeled planar graphs
f–j. the preceding five properties with plane graphs in place of planar graphs
k. plane graphs rooted by the following three-stage procedure:
    i.   Select an arbitrary edge x of a plane graph G.
    ii. Orient x arbitrarily.
    iii. Arbitrarily designate one of the two faces incident with x as the exterior.

There are q possibilities for step (i) since G has q lines, and two possibilities for each of the other two steps. Multiplying these together, we see that the total number of Tutte-orientations of a plane graph is $4q$, which is reflected in the paper [HT1] on the automorphism group of a planar graph. Tutte [T4] provided the definitive comprehensive survey of the art of enumerating Tutte-oriented plane maps, and we refer the reader to his work for all related information.

P7.1 *Planar and plane graphs*

All trees are planar, so the number of planar trees equals the number of trees. Plane unicyclic graphs are easily counted by using rooted plane trees for the figure counting series and taking the dihedral group as the configuration group in Pólya's Enumeration Theorem (2.4.6).

For both plane and planar graphs, all ten variations (a)–(j) above are open problems.

P7.2 *n-colorable graphs*

This has been solved only for $n = 2$, and is the same as problem P4.10.

P7.3 *n-colorable planar graphs*

This problem has not been solved for any $n > 1$. Even for $n = 2$, it is subtle because it involves homeomorphisms of $K_{3,3}$.

P7.4   *Self-dual plane graphs*

Given a plane graph G, its *dual* G^* is constructed as follows: Place a point in each region of G including the exterior region and if two regions have a line x in common, join the corresponding points by a line x^* crossing only x. The result is always a plane general graph in which loops and multiple lines are allowed. The tetrahedron is self-dual, while the cube and the octahedron are duals, as are the dodecahedron and the icosahedron.

The confrontation of manageable solutions to P7.1 and to P7.3 for $n = 4$ would settle the EE4CC and hence the 4CC itself. Of course this could also be accomplished by comparing any other pair of classes of planar graphs that are obtained in the same way, as for example by the same kind of rooting. It is safe to predict that the 4CC will not be settled for the first time by means of the EE4CC. Furthermore, the occasional rumors to the effect that the 4CC has been solved do not in any way answer these most difficult counting questions.

10.8   VARIATIONS ON GRAPHS

There are many configurations which are not graphs *per se* but are graphical in nature. Space permits only an indication of the counting problems for a few of these structures. These include simplicial complexes, Latin squares, knots, animals, and chessboard and paving configurations.

P8.1   *Simplicial complexes*

A *simplicial complex* consists of a finite nonempty set V of *points* and a collection of subsets of V called *simplexes* such that every point is a simplex and every nonnull subset of a simplex is one. The *dimension of a simplex S* is $|S| - 1$; the *dimension of a complex* is the maximum dimension of its simplexes.

While the number of n-dimensional complexes has only been solved for $n = 1$ when these become graphs (strictly speaking, graphs which are not totally disconnected), there is another special case which can be handled. In a *pure n-complex*, every maximal simplex has dimension n or 0. Thus every graph is a pure 1-complex. An indication of how to count pure n-complexes is given in [H11]. For example, the counting problem for pure 2-complexes with p points is obtained when $1 + x$ is substituted into the cycle index of the *triad group* $S_p^{(3)}$: this group is induced by S_p but acts on 3-subsets of objects. A formula for this cycle index appears in [O1]. Let $s(y) = \sum s_p y^p$ be the generating function for pure three-dimensional simplicial complexes.

We have found by brutal computation that this function begins

$$s(y) = y + y^2 + 2y^3 + 5y^4 + 34y^5 + 2136y^6 + 7\,013\,488y^5 + \cdots.$$

Now let s_{pqr} be the total number of complexes with dimension at most two with p points (0-simplexes), q lines (1-simplexes) and r cells (2-simplexes). Let $s_p(x, y) = \sum s_{pqr} a^q y^r$ be the counting polynomial for those of order p. For $p = 4$, this polynomial is

$$s_4(x, y) = 1 + x + 2x^2 + x^3(3 + y) + x^4(2 + y) + x^5(1 + y + y^2)$$
$$+ x^6(1 + y + y^2 + y^3 + y^4).$$

The problem is to determine $s_p(x, y)$ for arbitrary p. It should then be possible to derive the result for higher dimensional complexes.

P8.2   *Latin squares*

A *Latin square* is a square matrix of order n in which each row and column is a permutation of the integers $1, 2, \ldots, n$. If L_n is the number of squares of order n in which the first row and the first column are in standard order $1, 2, \ldots, n$, then the total number of latin squares of order n is obviously $n!(n - 1)!L_n$. The following values of L_n for $n \leq 7$ are taken from Riordan [R15, p. 210]:

$$x + x^2 + x^3 + 4x^4 + 56x^5 + 9408x^6 + 16\,942\,080x^7 + \cdots.$$

Latin squares may also be regarded as bicolored graphs $K_{n,n}$ in which the lines are also colored. The points u_i of the first color correspond to the rows of the latin square while the points v_i of the other color stand for the columns. Every line of $K_{n,n}$ is colored with one of the n colors so that each point is incident with one line of each color.

The number of Latin squares is reduced if we introduce an equivalence that allows permutations of the n symbols and which does not distinguish between rows, columns or symbols. On this basis, J. J. Seidel (unpublished notes) compiled the data in the polynomial

$$x + x^2 + x^3 + 2x^4 + 2x^5 + 12x^6 + 147x^7 + \cdots.$$

Let $A = [a_{ij}]$ and $B = [b_{ij}]$ be two $n \times n$ Latin squares. Then A and B are called *orthogonal* if the n^2 pairs (a_{ij}, b_{ij}) are all distinct. The number of orthogonal pairs of Latin squares, as well as orthogonal m-tuples is open.

P8.3   *Knots and knot-graphs*

The first seven knots in the table in Reidemeister [R11, p. 70], which were taken from the article by Alexander and Briggs [AB1], are shown in Figure

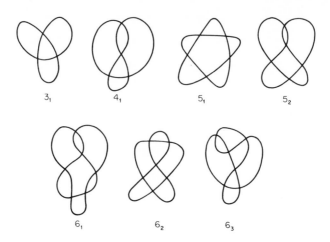

Figure 10.8.1

The smallest knots.

10.8.1. Each of these is a plane projection with a minimum number of crossings. To each such minimal plane projection of a knot, it is possible to associate a graph by taking the crossing points as the vertices and the arcs joining a pair of consecutive crossing points as the edges. Thus the graphs (actually multigraphs) of these seven knots are shown in Figure 10.8.2. Obviously each such graph is regular of degree 4. A *knot-graph* is the graph of a minimal plane projection of a knot. Let $G_{p,n}$ be the graph of the knot denoted in Figure 10.8.1 by p_n.

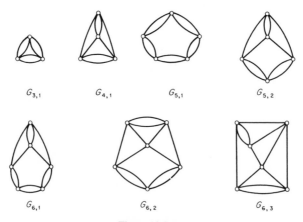

Figure 10.8.2

The smallest knot-graphs.

Figure 10.8.3

The knot-graph of four different knots.

As pointed out by Brown [B1], the graph of a knot is not a knot invariant. This he verified by presenting the "knot-product" of knots 3_1 and 4_1 (the definition of knot-product is reported in [H2]) in two different ways, so that one of the knot-graphs has four double edges and the other has three. Thus we must speak of *a* graph of a knot rather than *the* graph of a knot!

Conversely, different knots can have the same knot graph; the most striking known example being the graph of knots 8_{17}, 8_{19}, 8_{20}, and 8_{21}, all of which have the graph of Figure 10.8.3.

Brown then developed the following criterion:

A plane multigraph is a graph of a knot if and only if

a.   it is regular of degree 4;
b.   it is a block;
c.   the plane curve created by starting at any vertex and any of its edges and then joining the ends of opposite edges at each vertex is an eulerian trail.

Several enumeration questions suggest themselves naturally now that we know which graphs are knot graphs.

1.   How many knot-graphs are there with p vertices?

Note that this is a different question from asking for the number of knots in view of the fact noted above that the graph of a knot is not a knot invariant.

2.   How many knot graphs are there with p vertices and m pairs of multiple edges? It appears from the available data that always $m \geq 2$.
3.   On replacing each pair of double edges by a single edge, a genuine graph (rather than a multigraph) results. How many such graphs are there?
4.   Which special classes of knot graphs can be identified and enumerated?

P8.4 *Chessboard configurations*

We mention only a few of the many problems associated with laying pieces on an $n \times n$ chessboard. All these problems may be interpreted in terms of graphs, and can be lavishly illustrated.

The *rooks problem* asks for the number of ways in which n rooks can be positioned on an $n \times n$ board so that no rook can threaten another. Thus no two rooks are permitted to be in the same row or column. The answer is $n!$ unless we identify solutions with respect to rotations and reflections of the board that preserve the two colors of the squares. In the latter case the number of different configurations has been computed for $n \leq 7$ and the data, Kraitchik [K5], are given by

$$x + x^2 + 2x^3 + 7x^4 + 23x^5 + 115x^6 + 694x^7 + \cdots.$$

Any solution of the rooks problem in which no two pieces are on the same diagonal is clearly a solution of the corresponding *queens problem*. The latter is unsolved in both the labeled and unlabeled cases. The number of configurations with $n \leq 12$ has been computed in Kraitchik [K5] when color-preserving rotations and reflections of the board are allowed. The data are given by:

$$x + 2x^4 + 10x^5 + 4x^6 + 40x^7 + 92x^8 + 352x^9$$

$$+ 724x^{10} + 2680x^{11} + 14\,200x^{12} + \cdots.$$

Similar problems may be posed for bishops and knights. Variations can also be introduced by changing the shape of the board to rectangular, L shapes or triangles. A more interesting rooks problem from the standpoint of enumeration is to determine the number of distinct rook polynomials, defined in Riordan [R15, p. 165].

Another type of chessboard problem asks for the number of different tours the knight can make on an $m \times n$ board in which each square is visited exactly once. These are of course labeled hamiltonian cycles in a knightlike graph of order mn. Numerous partial results may be found in [K5].

P8.5 *Cell growth problems*

All of the "square animals" with at most four cells are shown in Figure 10.8.4. Thus an animal grows in the plane of adding a square cell of the same

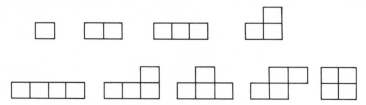

Figure 10.8.4
The smallest square animals.

Figure 10.8.5

The smallest holey animal.

size to any of its sides. Furthermore, animals are assumed to be simply connected in that they have no holes.

Read [R4] devised a clever scheme that enabled him to compute the number of animals, holey or not, with as many as ten cells. These results and those of Parkin (unpublished) led to Table 10.8.1. Klarner [K4] used elegant analytic methods to establish lower bounds for both kinds of animals, but no formulas have been found from which the number of animals with n cells can be calculated. Note that the smallest animal not simply connected, Figure 10.8.5, has seven cells and is the only "holey animal" of order 7. Thus with seven cells there are 107 square animals and just one holey animal.

TABLE 10.8.1

SQUARE ANIMALS

Cells	1	2	3	4	5	6	7	8	9	10
Animals	1	1	2	5	12	35	107	363	1 248	4 271
Holey animals	0	0	0	0	0	0	1	6	37	384

There are numerous variations of this problem [H12]. With triangles instead of squares as the basic cells, the desired series begins

$$x + x^2 + x^3 + 3x^4 + 4x^5 + \cdots.$$

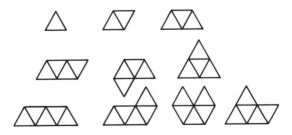

Figure 10.8.6

The smallest triangular animals.

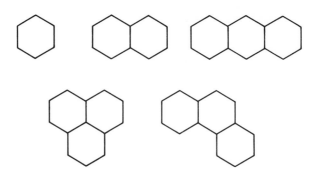

Figure 10.8.7

The smallest hexagonal animals.

With hexagons instead of squares we have

$$x + x^2 + 3x^3 + 7x^4 + 22x^5 + 83x^6 + \cdots.$$

We can ask for the number of toroidal animals in which the genus of the underlying surface is 1 instead of 0. Finally we can use cubes instead of squares and ask for the number of solid animals. Many other related questions involving paving problems may be found in Golomb's book [H3], which show vividly how such constructions lead to tantalizing puzzles.

P8.6   *Ising Problem*

Ising [I1] proposed the problem which bears his name and solved it for the one-dimensional case. Onsager [O2] was first to find a solution to the two-dimensional problem, but no progress has been made toward solving the problem in higher dimensions.

In the *two-dimensional lattice graph* $L_{m,n}$ the points are ordered pairs (i, j), $i = 1, 2, \ldots, m$; $j = 1, 2, \ldots, n$. Two points are adjacent if the Euclidean distance between them is 1. Thus $L_{m,n}$ is the cartesian product $P_m \times P_n$ of two paths. Frequently, in physical applications these graphs are drawn on the torus with opposite sides identified. A *d-dimensional lattice graph* is defined similarly. The problem is to determine the number A_q of different labeled even subgraphs with q lines.

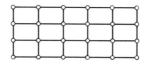

Figure 10.8.8

A lattice graph.

By the *area* of an even subgraph of a two-dimensional lattice we mean the minimum area enclosed by the line-disjoint cycles of this subgraph. The "two-dimensional Ising problem with magnetic field" asks for the generating function that counts labeled even subgraphs with both the number of lines and the area as enumeration parameters.

As another variation of the Ising problem, there is the case known in the literature as the "interaction between nonnearest neighbors." Consider the supergraph of a lattice graph obtained by adding both diagonals into each of its squares. The problem again is to count the even subgraphs of such a graph.

For a more detailed discussion of the various Ising problems, one may consult the expository article [H14].

Appendix I

This appendix presents nine tables that list the number of graphs of various kinds. Although much of this information is in the text, it is still convenient to gather these data together here. The tables and their contents are as follows. In the first two tables the parameters are p points and q lines; in all the others, just p points.

Table
A1 Graphs
A2 Connected graphs
A3 Graphs, connected graphs, and blocks
A4 Digraphs, connected digraphs, and symmetric relations
A5 Self-complementary digraphs and self-converse digraphs and relations
A6 Labeled finite topologies
A7 Trees, rooted trees, identity trees, and homeomorphically irreducible trees
A8 Tournaments
A9 Asymptotic approximation to graphs and tournaments.

The format of this table for the number of (p, q) graphs follows that in [H1, p. 214], the main difference being that there are no misprints here. The table first appeared in [R15, p. 146].

TABLE A1

THE NUMBER OF (p, q) GRAPHS

q \ p	1	2	3	4	5	6	7	8	9
0	1	1	1	1	1	1	1	1	1
1		1	1	1	1	1	1	1	1
2			1	2	2	2	2	2	2
3			1	3	4	5	5	5	5
4				2	6	9	10	11	11
5				1	6	15	21	24	25
6				1	6	21	41	56	63
7					4	24	65	115	148
8					2	24	97	221	345
9					1	21	131	402	771
10					1	15	148	663	1 637
11						9	148	980	3 252
12						5	131	1 312	5 995
13						2	97	1 557	10 120
14						1	65	1 646	15 615
15						1	41	1 557	21 933
16							21	1 312	27 987
17							10	980	32 403
18							5	663	34 040
g_p	1	2	4	11	34	156	1 044	12 346	274 668

The entries for the number of connected graphs were taken from Cadogan [C1]. A few more can be obtained from Sloane's catalogue [S4].

TABLE A2

THE NUMBER OF CONNECTED (p, q) GRAPHS

p \ q	0	1	2	3	4	5	6	7	8	9	10	11	12	13
1	1													
2		1												
3			1	1										
4				2	2	1	1							
5					3	5	5	4	2	1	1			
6						6	13	19	22	20	14	9	5	2
7							11	33	67	107	132	138	126	95
8								23	89	236	486	814	1 169	1 454

The numbers (Table A3, next page) g_p, $p \le 9$, appear in Riordan [R15, p. 146], [H1, p. 214], Oberschelp [O1], Robinson [R19], and also in other references. King and Palmer [KP1] used a computer to calculate g_p for $p \le 24$. Cadogan [C1] calculated c_p for $p \le 8$, and Robinson [R19] worked out b_p for $p \le 9$.

The numbers d_p of digraphs were computed by Oberschelp [O1], as were the symmetric relations for $p \le 8$. Sloane [S4] lists d_p for $p \le 11$.

TABLE A4

DIGRAPHS

p	Digraphs	Connected digraphs	Symmetric relations
1	1	1	2
2	3	2	6
3	16	13	20
4	218	199	90
5	9 608	9 364	544
6	1 540 944	1 530 843	5 096
7	882 033 440	880 471 142	79 264
8	1 793 359 192 848	1 792 473 955 306	2 208 612

TABLE A3

GRAPHS

p	graphs g_p	connected graphs c_p	blocks b_p
1	1	1	0
2	2	1	1
3	4	2	1
4	11	6	3
5	34	21	10
6	156	112	56
7	1044	853	468
8	12 346	11 117	7 123
9	274 668	261 080	194 066
10	12 005 168	11 716 571	
11	1 018 997 864		
12	165 091 172 592		
13	50 502 031 367 952		
14	29 054 155 657 235 488		
15	31 426 485 969 804 308 768		
16	64 001 015 704 527 557 894 928		
17	245 935 864 153 532 932 683 719 776		
18	1 787 577 725 145 611 700 547 878 190 848		
19	24 637 809 253 125 004 524 383 007 491 432 768		
20	645 490 122 795 799 841 856 164 638 490 742 749 440		
21	32 220 272 899 808 983 433 502 244 253 755 283 616 097 664		
22	3 070 846 483 094 144 300 637 568 517 187 105 410 586 657 814 272		
23	559 946 939 699 792 080 597 976 380 819 462 179 812 276 348 458 981 632		
24	195 704 906 302 078 447 922 174 862 416 726 256 004 122 075 267 063 365 754 368		

Self-dual digraphs with respect to complementation and conversion, as well as self-converse relations are now tabulated.

TABLE A5

SELF-DUAL RELATIONS

p	Self-complementary digraphs	Self-converse digraphs	Self-converse relations
1	1	1	2
2	1	3	8
3	4	10	44
4	10	70	436
5	136	708	7 176
6	720	15 248	222 368
7	44 224	543 520	
8	703 760		

The number of labeled finite topologies on p points, which is equal to the number of labeled transitive digraphs, has been calculated for $p \le 7$ in Evans, Harary, and Lynn [EHL1]. B. Stubblefield has determined this number for $n = 8$.

TABLE A6

LABELED FINITE TOPOLOGIES

p	Labeled topologies
1	1
2	4
3	29
4	355
5	6 942
6	209 527
7	9 535 241
8	642 779 354

For the number of trees, rooted trees, identity trees, and homeomorphically irreducible trees, we have appropriated the table in [H1, p. 232], which in turn was taken in part from Riordan [R15, p. 138]. Allen Schwenk at the University of Michigan has used a computer to calculate all these parameters for $p \leq 39$.

TABLE A7

TREES

p	t_p	T_p	i_p	h_p	p	t_p	T_p
1	1	1	1	1	13	1 301	12 486
2	1	1	0	1	14	3 159	32 973
3	1	2	0	0	15	7 741	87 811
4	2	4	0	1	16	19 320	235 381
5	3	9	0	1	17	48 629	634 847
6	6	20	0	2	18	123 867	1 721 159
7	11	48	1	2	19	317 955	4 688 676
8	23	115	1	4	20	823 065	12 826 228
9	47	286	3	5	21	2 144 505	35 221 832
10	106	719	6	10	22	5 623 756	97 055 181
11	235	1 842	15	14	23	14 828 074	268 282 855
12	551	4 766	29	26	24	39 299 897	743 724 984
					25	104 636 890	2 067 174 645
					26	279 793 450	5 759 636 510

Paul Stein programmed the computer at the Los Alamos Scientific Laboratory to obtain the number of tournaments for $p \leq 30$, Table A8, next page.

In Table A9, p.246, the entries give $2^{p(p-1)/2}/p!$ which we saw in Chapter 9 gives an asymptotic approximation for both the number g_p of graphs and the number $T(p)$ of tournaments with p points. Again these numbers were kindly provided by Paul Stein.

TABLE A8

TOURNAMENTS

p	$T(p)$
1	1
2	1
3	2
4	4
5	12
6	56
7	456
8	6 880
9	191 536
10	9 733 056
11	903 753 248
12	154 108 311 168
13	48 542 114 686 912
14	28 401 423 719 122 304
15	31 021 002 160 355 166 848
16	63 530 415 842 308 265 100 288
17	244 912 778 438 520 759 443 245 824
18	1 783 398 846 284 777 975 419 600 287 232
19	24 605 641 171 260 376 770 598 003 978 281 472
20	645 022 068 557 873 570 931 850 526 424 042 500 096
21	32 207 364 031 661 175 384 456 332 260 036 660 040 346 624
22	3 070 169 883 150 468 336 193 188 889 176 239 554 269 865 953 280
23	559 879 382 429 394 075 397 997 876 821 117 309 031 348 506 639 435 776
24	1 956 920 276 575 218 760 843 168 426 608 334 827 851 734 377 775 365 039 898 624
25	131 326 696 677 895 002 131 450 257 709 457 767 557 170 027 052 967 027 982 788 816 896
26	169 484 335 125 246 268 100 514 597 385 576 342 667 201 246 238 506 672 327 765 919 863 947 264
27	421 255 599 848 131 447 082 003 884 098 323 929 861 369 544 621 589 389 269 735 693 986 231 100 612 608
28	2 019 284 625 667 208 265 086 928 694 043 799 677 058 780 746 074 756 618 649 807 453 554 008 410 636 526 845 952
29	18 691 296 182 213 712 407 784 892 577 100 643 237 772 159 079 535 345 610 331 272 616 359 410 643 727 554 822 061 146 112
30	334 493 774 260 141 796 028 606 267 674 709 437 232 608 940 215 918 926 763 659 414 050 175 507 824 571 200 950 884 097 540 096 000

TABLE A9

Asymptotic Number of Tournaments and Graphs

p	$2^{\binom{p}{2}}/p!$
1	1
2	1
3	1
4	2
5	8
6	45
7	416
8	6 657
9	189 372
10	9 695 869
11	902 597 327
12	154 043 277 297
13	48 535 481 831 642
14	28 400 190 511 772 276
15	31 020 581 422 991 798 557
16	63 530 150 754 287 203 445 810
17	244 912 468 225 468 597 942 626 507
18	1 783 398 168 624 923 337 196 441 201 196
19	24 605 638 395 579 573 858 211 783 276 124 626
20	645 022 047 157 081 180 948 706 971 513 641 417 725
21	32 207 363 719 989 693 161 641 493 398 185 145 868 257 184
22	3 070 169 874 550 173 863 332 853 689 226 853 410 359 422 313 750
23	559 879 381 978 490 975 464 019 198 266 910 789 847 137 671 663 161 697
24	195 692 027 612 492 717 696 053 131 613 974 749 458 250 743 733 957 146 609 599
25	131 326 696 669 310 184 928 548 229 462 563 319 608 278 228 391 409 823 336 516 656 557
26	169 484 335 122 115 195 657 399 862 470 076 055 903 470 709 678 462 703 895 275 433 435 498 631
27	421 255 599 845 942 668 803 919 924 597 149 559 454 904 184 585 016 179 214 085 529 813 478 905 185 546
28	2 019 284 625 664 270 536 611 378 920 477 168 898 079 934 218 310 482 029 759 803 193 187 192 943 926 121 612 944
29	18 691 296 182 206 129 806 987 592 941 623 330 025 693 264 356 626 820 456 495 804 849 429 663 146 992 789 154 562 961 044
30	334 493 774 260 104 102 715 593 766 508 469 331 712 364 208 913 311 144 737 971 969 122 910 864 452 626 959 027 797 528 575 878 038

Appendix II

The first table in this appendix summarizes the frequencies of the digraph groups for digraphs of order 4, and corrects four entries of the table in [HP6] where this data first appeared.

TABLE A10

ENUMERATION OF FOUR-POINT DIGRAPHS BY AUTOMORPHISM GROUP

Group	0	1	2	3	4	5	6	≥ 7	Total
S_4	1	0	0	0	0	0	0	1	2
$E_2 S_2$	0	1	2	3	6	10	6	22	50
E_4	0	0	1	7	16	28	32	52	136
$S_2 S_2$	0	0	1	0	1	0	2	2	6
$S_2[E_2]$	0	0	1	0	2	0	4	3	10
$E_1 S_3$	0	0	0	2	0	0	2	2	6
$E_1 C_3$	0	0	0	1	0	0	2	1	4
C_4	0	0	0	0	1	0	0	1	2
D_4	0	0	0	0	1	0	0	1	2
Total	1	1	5	13	27	38	48	85	218

The next table displays the data obtained by Stockmeyer [S6] for digraphs of order 5. As mentioned in problem P3.4 of Chapter 10, the calculation of these numbers requires the knowledge of the entire lattice of subgroups of S_5.

TABLE A11

ENUMERATION OF FIVE-POINT DIGRAPHS BY AUTOMORPHISM GROUP

Group \ Lines	0	1	2	3	4	5	6	7	8	9	10	≥ 11	Total
E_5	0	0	0	5	28	107	278	591	962	1 314	1 431	3 285	8 001
E_3S_2	0	0	1	8	13	43	59	105	124	168	148	521	1 190
$S_2[E_2]E_1$	0	0	1	0	6	0	22	0	38	0	49	67	183
E_2C_3	0	0	0	0	1	0	2	4	2	2	6	11	28
E_1C_4	0	0	0	0	1	0	0	0	3	0	0	4	8
$E_1S_2S_2$	0	0	2	0	7	0	13	0	18	0	26	40	106
C_5	0	0	0	0	0	1	0	0	0	0	1	1	3
E_2S_3	0	1	0	2	2	2	2	7	1	4	8	21	50
C_2C_3	0	0	0	1	0	1	0	0	0	2	0	4	8
E_1D_4	0	0	0	0	1	0	0	0	3	0	0	4	8
D_5	0	0	0	0	0	0	0	0	0	0	1	0	1
S_2S_3	0	0	1	0	0	0	3	0	3	0	0	7	14
E_1S_4	0	0	0	0	2	0	0	0	1	0	0	3	6
S_5	1	0	0	0	0	0	0	0	0	0	0	1	2
Total	1	1	5	16	61	154	379	707	1155	1490	1670	3969	9608

Appendix III

CYCLE INDEX FORMULAS FOR THE SYMMETRIC GROUPS S_n WITH $n \leqslant 10$

$$Z(S_0) = 1$$

$$Z(S_1) = s_1$$

$$Z(S_2) = \frac{1}{2!}(s_1^2 + s_2)$$

$$Z(S_3) = \frac{1}{3!}(s_1^3 + 3s_1s_2 + 2s_3)$$

$$Z(S_4) = \frac{1}{4!}(s_1^4 + 6s_1^2s_2 + 8s_1s_3 + 3s_2^2 + 6s_4)$$

$$Z(S_5) = \frac{1}{5!}(s_1^5 + 10s_1^3s_2 + 20s_1^2s_3 + 15s_1s_2^2 + 30s_1s_4 + 20s_2s_3 + 24s_5)$$

$$Z(S_6) = \frac{1}{6!}(s_1^6 + 15s_1^4s_2 + 40s_1^3s_3 + 45s_1^2s_2^2 + 90s_1^2s_4 + 120s_1s_2s_3$$

$$+ 144s_1s_5 + 15s_2^3 + 90s_2s_4 + 40s_3^2 + 120s_6).$$

$$Z(S_7) = \frac{1}{7!}(s_1^7 + 21s_1^5s_2 + 70s_1^4s_3 + 105s_1^3s_2^2 + 210s_1^3s_4 + 420s_1^2s_2s_3$$

$$+ 504s_1^2s_5 + 105s_1s_2^3 + 630s_1s_2s_4 + 280s_1s_3^2$$

$$+ 840s_1s_6 + 210s_2^2s_3 + 504s_2s_5 + 420s_3s_4 + 720s_7)$$

$$Z(S_8) = \frac{1}{8!}(s_1^8 + 28s_1^6s_2 + 112s_1^5s_3 + 210s_1^4s_2^2 + 420s_1^4s_4 + 1120s_1^3s_2s_3$$

$$+ 1344s_1^3s_5 + 420s_1^2s_2^3 + 2520s_1^2s_2s_4 + 1120s_1^2s_3^2 + 3360s_1^2s_6$$

$$+ 1680s_1s_2^2s_3 + 4032s_1s_2s_5 + 3360s_1s_3s_4 + 5760s_1s_7 + 105s_2^4$$

$$+ 1260s_2^2s_4 + 1120s_2s_3^2 + 3360s_2s_6 + 2688s_3s_5 + 1260s_4^2$$

$$+ 5040s_8)$$

$$Z(S_9) = \frac{1}{9!}(s_1^9 + 36s_1^7s_2 + 168s_1^6s_3 + 378s_1^5s_2^2 + 756s_1^5s_4 + 2520s_1^4s_2s_3$$

$$+ 3024s_1^4s_5 + 1260s_1^3s_2^3 + 7560s_1^3s_2s_4 + 3360s_1^3s_3^2 + 7560s_1^2s_2^2s_3$$

$$+ 945s_1s_2^4 + 10080s_1^3s_6 + 18144s_1^2s_2s_5 + 15120s_1^2s_3s_4$$

$$+ 25920s_1^2s_7 + 11340s_1s_2^2s_4 + 10080s_1s_2s_3^2 + 30240s_1s_2s_6$$

$$+ 24192s_1s_3s_5 + 11340s_1s_4^2 + 45360s_1s_8 + 2520s_2^3s_3$$

$$+ 9072s_2^2s_5 + 15120s_2s_3s_4 + 25920s_2s_7 + 2240s_3^3 + 20160s_3s_6$$

$$+ 18144s_4s_5 + 40320s_9)$$

$$Z(S_{10}) = \frac{1}{10!}(s_1^{10} + 45s_1^8s_2 + 240s_1^7s_3 + 630s_1^6s_2^2 + 1260s_1^6s_4 + 5040s_1^5s_2s_3$$

$$+ 6048s_1^5s_5 + 3150s_1^4s_2^3 + 18900s_1^4s_2s_4 + 8400s_1^4s_3^2$$

$$+ 25200s_1^4s_6 + 25200s_1^3s_2^2s_3 + 60480s_1^3s_2s_5 + 50400s_1^3s_3s_4$$

$$+ 86400s_1^3s_7 + 4725s_1^2s_2^4 + 56700s_1^2s_2^2s_4 + 50400s_1^2s_2s_3^2$$

$$+ 151200s_1^2s_2s_6 + 120960s_1^2s_3s_5 + 56700s_1^2s_4^2 + 226800s_1^2s_8$$

$$+ 25200s_1s_2^3s_3 + 90720s_1s_2^2s_5 + 151200s_1s_2s_3s_4 + 259200s_1s_2s_7$$

$$+ 22400s_1s_3^3 + 201600s_1s_3s_6 + 181440s_1s_4s_5 + 403200s_1s_9$$

$$+ 945s_2^5 + 18900s_2^3s_4 + 25200s_2^2s_3^2 + 75600s_2^2s_6 + 120960s_2s_3s_5$$

$$+ 56700s_2s_4^2 + 226800s_2s_8 + 50400s_3^2s_4 + 172800s_3s_7$$

$$+ 151200s_4s_6 + 72576s_5^2 + 362880s_{10})$$

CYCLE INDEX FORMULAS FOR THE PAIR GROUPS $S_n^{(2)}$ WITH $n \leqslant 10$

$Z(S_2^{(2)}) = s_1$

$Z(S_3^{(2)}) = \dfrac{1}{3!}(s_1^3 + 3s_1s_2 + 2s_3).$

$Z(S_4^{(2)}) = \dfrac{1}{4!}(s_1^6 + 6s_1^2s_2^2 + 8s_3^2 + 3s_1^2s_2^2 + 6s_2s_4).$

$Z(S_5^{(2)}) = \dfrac{1}{5!}(s_1^{10} + 10s_1^4s_2^3 + 20s_1s_3^3 + 15s_1^2s_2^4 + 30s_2s_4^2 + 20s_1s_3s_6 + 24s_5^2).$

$Z(S_6^{(2)}) = \dfrac{1}{6!}(s_1^{15} + 15s_1^7s_2^4 + 40s_1^3s_3^4 + 45s_1^3s_2^6 + 90s_1s_2s_4^3 + 120s_1s_2s_3^2s_6$
$\qquad\qquad + 144s_5^3 + 15s_1^3s_2^6 + 90s_1s_2s_4^3 + 40s_3^5 + 120s_3s_6^2).$

$Z(S_7^{(2)}) = \dfrac{1}{7!}(s_1^{21} + 21s_1^{11}s_2^5 + 70s_1^6s_3^5 + 105s_1^5s_2^8 + 210s_1^3s_2s_4^4$

$\qquad\qquad + 420s_1^2s_2^2s_3^3s_6 + 504s_1s_5^4 + 105s_1^3s_2^9 + 630s_1s_2^2s_4^4 + 280s_3^7$

$\qquad\qquad + 840s_3s_6^3 + 210s_1^2s_2^2s_3s_6^2 + 504s_1s_5^2s_{10}$

$\qquad\qquad + 420s_2s_3s_4s_{12} + 720s_7^3).$

$Z(S_8^{(2)}) = \dfrac{1}{8!}(s_1^{28} + 28s_1^{16}s_2^6 + 112s_1^{10}s_3^6 + 210s_1^8s_2^{10} + 420s_1^6s_2s_4^5$

$\qquad\qquad + 1120s_1^4s_2^3s_3^4s_6 + 1344s_1^3s_3^5 + 420s_1^4s_2^{12} + 2520s_1^2s_2^3s_4^5$

$\qquad\qquad + 1120s_1s_3^9 + 3360s_1s_3s_6^4 + 1680s_1^2s_2^4s_3^2s_6 + 4032s_1s_2s_5^3s_{10}$

$\qquad\qquad + 3360s_2s_3^2s_4^2s_{12} + 5760s_7^4 + 105s_1^4s_2^{12} + 1260s_1^2s_2^3s_4^5$

$\qquad\qquad + 1120s_1s_3^5s_6^2 + 3360s_1s_3s_6^4 + 2688s_3s_5^2s_{15} + 1260s_2^2s_4^6$

$\qquad\qquad + 5040s_4s_8^3).$

$$Z(S_9^{(2)}) = \frac{1}{9!}(s_1^{36} + 36s_1^{22}s_2^7 + 168s_1^{15}s_3^7 + 378s_1^{12}s_2^{12} + 756s_1^{10}s_2s_4^6$$

$$+ 2520_1^7s_2^4s_3^5s_6 + 3024s_1^6s_5^6 + 1260s_1^6s_2^{15} + 7560s_1^4s_2^4s_4^6$$

$$+ 3360s_1^3s_3^{11} + 7560s_1^3s_2^6s_3^3s_6^2 + 945s_1^4s_2^{16} + 10080s_1^3s_3s_6^5$$

$$+ 18144s_1^2s_2^2s_5^4s_{10} + 15120s_1s_2s_3^3s_4^3s_{12} + 25920s_1s_7^5$$

$$+ 11340s_1^2s_2^5s_4^6 + 10080s_1s_2s_3^7s_6^2 + 30240s_1s_2s_3s_6^5 + 24192s_3^2s_5^3s_{15}$$

$$+ 11340s_2^2s_4^8 + 45360s_4s_8^4 + 2520s_1^3s_2^6s_3s_6^3 + 9072s_1^2s_2^2s_5^2s_{10}^2$$

$$+ 15120s_1s_2s_3s_4^3s_6s_{12} + 25920s_1s_7^3s_{14} + 2240s_3^{12} + 20160s_3^3s_6^5$$

$$+ 18144s_2s_4s_5^2s_{20} + 40320s_9^4).$$

$$Z(S_{10}^{(2)}) = \frac{1}{10!}(s_1^{45} + 45s_1^{29}s_2^8 + 240s_1^{21}s_3^8 + 630s_1^{17}s_2^{14} + 1260s_1^{15}s_2s_4^7$$

$$+ 5040s_1^{11}s_2^5s_3^6s_6 + 6048s_1^{10}s_5^7 + 3150s_1^9s_2^{18} + 18900s_1^7s_2^5s_4^7$$

$$+ 8400s_1^6s_3^{13} + 25200s_1^6s_3s_6^6 + 25200s_1^5s_2^8s_3^4s_6^2 + 60480s_1^4s_2^3s_5^5s_{10}$$

$$+ 50400s_1^3s_2s_3^4s_4^4s_{12} + 86400s_1^3s_7^6 + 4725s_1^5s_2^{20} + 56700s_1^3s_2^7s_4^7$$

$$+ 50400s_1^2s_2^2s_3^9s_6^2 + 151200s_1^2s_2^2s_3s_6^6 + 120960s_1s_3^3s_5^4s_{15}$$

$$+ 56700s_1s_2^2s_4^{10} + 226800s_1s_4s_8^5 + 25200s_1^3s_2^9s_3^2s_6^3$$

$$+ 90720s_1^2s_2^4s_5^3s_{10}^2 + 151200s_1s_2^2s_3^2s_4^4s_6s_{12} + 259200s_1s_2s_7^4s_{14}$$

$$+ 22400s_3^{15} + 201600s_3^3s_6^6 + 181440s_2s_4^2s_5^3s_{20} + 403200s_9^5$$

$$+ 945s_1^5s_2^{20} + 18900s_1^3s_2^7s_4^7 + 25200s_1^2s_2^2s_3^5s_6^4 + 75600s_1^2s_2^2s_3s_6^6$$

$$+ 120960s_1s_3s_5^2s_6s_{10}s_{15} + 56700s_1s_2^2s_4^{10} + 226800s_1s_4s_8^5$$

$$+ 50400s_2s_3^5s_4s_{12}^2 + 172800s_3s_7^3s_{21} + 151200s_2s_3s_4s_6^2s_{12}^2$$

$$+ 72576s_5^9 + 362880s_5s_{10}^4).$$

It is the custom to ornament every scientific work with a bibliography—to prove the author's competence by showing the mountain of dross he has sifted to win one nugget of truth.

Dr. Lawrence J. Peter, hierarchiologist

BIBLIOGRAPHY

Alexander, J. W.
[AB1]   (with G. B. Briggs). On types of knotted curves, *Ann. of Math.* **28**, 562–586 (1938).
Austin, T. L.
[A1]   The enumeration of point labelled chromatic graphs and trees, *Canad. J. Math.* **12**, 535–545 (1960).
Beineke, L. W.
[BM1]   (with J. W. Moon). Several proofs of the number of labeled 2-dimensional trees, *in* "Proof Techniques in Graph Theory" (F. Harary, ed.), pp. 11–20. Academic Press, New York, 1969.
[BP1]   (with R. E. Pippert). On the number of k-dimensional trees, *J. Combinatorial Theory* **6**, 200–205 (1969).
Bott, R.
[BM2]   (with J. P. Mayberry). Matrices and trees, *in* "Economic Activity Analysis" (O. Morgenstern, ed.), pp. 391–400. Wiley, New York, 1954.
Briggs, G. B.
        See [AB1].
Brooks, R. L.
[BSST1] (with C. A. B. Smith, A. H. Stone, and W. T. Tutte). The dissection of rectangles into squares, *Duke Math. J.* **7**, 312–340 (1940).

Brown, T. A.
[B1]   A note on some graphs related to knots, *J. Combinatorial Theory* **1**, 498–502 (1966).
Brown, W. G.
[B2]   Enumeration of triangulations of the disk, *Proc. London Math. Soc.* **14**, 746–768 (1964).
[B3]   Historical note on a recurrent combinatorial problem, *Amer. Math. Monthly* **72**, 973–977 (1965).
deBruin, N. G.
[B4]   A combinatorial problem, *Indag. Math.* **8**, 461–467 (1946).
[B5]   Pólya's theory of counting, *in* "Applied Combinatorial Mathematics" (E. F. Beckenbach, ed.), pp. 144–184. Wiley, New York, 1964.
[BE1]   (with T. van Aardenne Ehrenfest). Circuits and trees in oriented graphs, *Simon Stevin* **28**, 203–217 (1951).
Burnside, W.
[B7]   "Theory of Groups of Finite Order," 2nd ed., p. 191, Theorem VII. Cambridge Univ. Press, London, 1911. Reprinted by Dover, New York, 1955.
Cadogan, C. C.
[C1]   The möbius function and connected graphs, *J. Combinatorial Theory*, **11B**, 193–200 (1971).
Cartwright, D.
[CG1]   (with T. Gleason). The number of paths and cycles in a digraph, *Psychometrika* **31**, 179–199 (1966). See also [HNC1].
Cayley, A.
[C2]   A theorem on trees, *Quart. J. Math. Oxford Ser.* **23**, 376–378 (1889); *Collected Papers*, Cambridge **13**, 26–28 (1897).
Chao, C. Y.
[C3]   On the classification of symmetric graphs with a prime number of vertices, *Trans. Amer. Math. Soc.* **158**, 247–256 (1971).
Chvátal, V.
[CH1]   (with F. Harary). Generalized Ramsey theory for graphs, *Bull. Amer. Math. Soc.* (1972).
Clifford, W. K.
[C4]   "Mathematical Papers," pp. 1–16. London, 1882.
Comtet, L.
[C5]   "Analyse Combinatoire," Vol. I. Presses Universitaires de France, Paris, 1970.
Davis, R. L.
[D1]   The number of structures of finite relations, *Proc. Amer. Math. Soc.* **4**, 486–495 (1953).
Douglas, R. J.
[D2]   Tournaments that admit exactly one Hamiltonian cycle, *Proc. London Math. Soc.* **21**, 716–730 (1970).
Ehrenfest, T. van Aardenne
         See [BE1]
Erdös, P.
[ER1]   (with A. Rényi). Asymmetric graphs, *Acta. Math. Acad. Sci. Hungar.* **14**, 293–315 (1963).
Euler, L.
[E1]   *Novi Comment. Acad. Sci. Imperialis Petropolitanae* **7**, 13–14 (1758–1759).

Evans, J. W.
[EHL1]   (with F. Harary and M. S. Lynn). On the computer enumeration of finite topologies, *Comm. ACM* **10**, 295–298 (1967).

Ford, G. W.
[FNU1]   (with R. Z. Norman and G. E. Uhlenbeck). Combinatorial problems in the theory of graphs II, *Proc. Nat. Acad. Sci. U.S.A.* **42**, 203–208 (1956).
[FU1]   (with G. E. Uhlenbeck). Combinatorial problems in the theory of graphs I, III, and IV, *Proc. Nat. Acad. Sci. U.S.A.* **42**, 122–128, 529–535 (1956); **43**, 163–167 (1957).
         See [UF1].

Foulkes, H. O.
[F1]   On Redfield's group reduction functions, *Canad. J. Math.*, **15**, 272–284 (1963).
[F2]   On Redfield's range-correspondences, *Canad. J. Math.* **18**, 1060–1071 (1966).

Frucht, R.
[FH1]   (with F. Harary). Self-complementary generalized orbits of a permutation group, *Canad. Math. Bull.* to be published.

Geller, D. P.
[G1]   The square root of a digraph, *J. Combinatorial Theory* **5**, 320–322 (1968).

Gilbert, E. N.
[G2]   Enumeration of labelled graphs, *Canad. J. Math.* **8**, 405–411 (1956).

Gleason, T.
         See [CG1].

Golomb, S.
[G3]   "Polyominoes." Scribner's, New York, 1965.

Goursat, E.
[GH1]   (with E. R. Hedrick). "A Course in Mathematical Analysis." Vol. 1, pp. 404–405. Ginn, Boston, 1904.

Harary, F.
[H1]   "Graph Theory." Addison-Wesley, Reading, Massachusetts, 1969.
[H2]   On the algebraic structure of knots, *Amer. Math. Monthly* **56**, 466–468 (1949).
[H3]   On the notion of balance of a signed graph, *Michigan Math. J.* **2**, 143–146 (1953–1954).
[H4]   The number of linear, directed, rooted, and connected graphs, *Trans. Amer. Math. Soc.* **78**, 445–463 (1955).
[H5]   The number of dissimilar supergraphs of a linear graph, *Pacific J. Math.* **7**, 903–911 (1957).
[H6]   The number of oriented graphs, *Michigan Math. J.* **4**, 221–224 (1957).
[H7]   On the number of bicolored graphs, *Pacific J. Math.* **8**, 743–755 (1958).
[H8]   Note on Carnap's relational asymptotic relative frequencies, *J. Symbolic Logic* **23**, 257–260 (1958).
[H9]   Exponentiation of permutation groups, *Amer. Math. Monthly* **66**, 572–575 (1959).
[H10]   The number of functional digraphs, *Math. Ann.* **138**, 203–210 (1959).
[H11]   Applications of Pólya's theorem to permutation groups, *in* "A Seminar on Graph Theory" (F. Harary, ed.), pp. 25–33. Holt, New York, 1967.
[H12]   Graphical enumeration problems, *in* "Graph Theory and Theoretical Physics" (F. Harary, ed.), pp. 1–41. Academic Press, New York, 1967.
[H13]   Covering and packing in graphs I, *Ann. N.Y. Acad. Sci.* **175**, 198–205 (1970).
[H14]   A graphical exposition of the Ising problem, *J. Austral. Math. Soc.* **12**, 365–377 (1971).

[HKMR1]   (with C. King, A. Mowshowitz, and R. C. Read). Cospectral graphs and digraphs, *Bull. London Math. Soc.* **3**, 321–328 (1971).

[HM1]   (with B. Manvel). On the number of cycles in a graph, *Mat. Časopis Sloven. Akad. Vied.* **21**, 55–63 (1971).

[HMR1]   (with A. Mowshowitz and J. Riordan). Labeled trees with unlabeled endpoints, *J. Combinatorial Theory* **6**, 60–64 (1969).

[HN1]   (with R. Z. Norman). The dissimilarity characteristic of Husimi trees, *Ann. of Math.* **58**, 134–141 (1953).

[HN2]   (with R. Z. Norman). Dissimilarity characteristic theorems for graphs, *Proc. Amer. Math. Soc.* **11**, 332–334 (1960).

[HN3]   (with R. Z. Norman). Some properties of line digraphs, *Rend. Circ. Mat. Palermo* **9**, 161–168 (1961).

[HNC1]   (with R. Z. Norman, and D. Cartwright). "Structural Models: An Introduction to the Theory of Directed Graphs." Wiley, New York, 1965.

[HP1]   (with E. M. Palmer), The number of graphs rooted at an oriented line, *ICC Bull.* **4**, 91–98 (1965).

[HP2]   (with E. M. Palmer). On the number of orientations of a given graph, *Bull. Acad. Polon. Sci. Ser. Sci. Math. Astronom. Phys.* **14**, 125–128 (1966).

[HP3]   (with E. M. Palmer). Enumeration of locally restricted digraphs, *Canad. J. Math.* **18**, 853–860 (1966).

[HP4]   (with E. M. Palmer). The power group enumeration theorem, *J. Combinatorial Theory* **1**, 157–173 (1966).

[HP5]   (with E. M. Palmer). Enumeration of self-converse digraphs, *Mathematika* **13**, 151–157 (1966).

[HP6]   (with E. M. Palmer). The groups of the small digraphs, *J. Indian Statist. Assoc.* **4**, 155–169 (1966).

[HP7]   (with E. M. Palmer). Enumeration of mixed graphs, *Proc. Amer. Math. Soc.* **17**, 682–687 (1966).

[HP8]   (with E. M. Palmer). The enumeration methods of Redfield, *Amer. J. Math.* **89**, 373–384 (1967).

[HP9]   (with E. M. Palmer). Enumeration of finite automata, *Information and Control* **10**, 499–508 (1967).

[HP10]   (with E. M. Palmer). On the number of balanced signed graphs, *Bull. Math. Biophys.* **29**, 759–765 (1967).

[HP11]   (with E. M. Palmer). Note on the number of forests. *Mat. Časopis Sloven. Akad. Vied.* **19**, 110–112 (1969).

[HP12]   (with E. M. Palmer). On acyclic simplicial complexes, *Mathematika* **15**, 119–122 (1968).

[HPR1]   (with E. M. Palmer and R. C. Read). The number of ways to label a structure, *Psychometrika* **32**, 155–156 (1967).

[HP14]   (with G. Prins). The number of homeomorphically irreducible trees, and other species, *Acta. Math.* **101**, 141–162 (1959).

[HP15]   (with G. Prins). Enumeration of bicolourable graphs, *Canad. J. Math.* **15**, 237–248 (1963).

[HP16]   (with G. Prins). The block-cutpoint-tree of a graph, *Publ. Math. Debrecen* **13**, 103–107 (1966).

[HPT1]   (with G. Prins and W. T. Tutte). The number of plane trees, *Indag. Math.* **26**, 319–329 (1964).

[HR1]   (with I. C. Ross). On the determination of redundancies in sociometric chains, *Psychometrika* **17**, 195–208 (1952).

[HT1]   (with W. T. Tutte). On the order of the group of a planar map, *J. Combinatorial Theory* **1**, 394–395 (1966).

[HU1]   (with G. E. Uhlenbeck). On the number of Husimi trees, *Proc. Nat. Acad. Sci. U.S.A.* **39**, 315–322 (1953).
          See also [CH1], [EHL1], and [FH1].

Harrison, M. H.

[H15]   A census of finite automata, *Canad. J. Math.* **17**, 100–113 (1965).

[HH1]   (with R. G. High). On the cycle index of a product of permutation groups, *J. Combinatorial Theory* **4**, 277–299 (1968).

Hedrick, E. R.
          See [GH1].

High, R. G.
          See [HH1].

Husimi, K.

[H16]   Note on Mayer's theory of cluster integrals, *J. Chem. Phys.* **18**, 682–684 (1950).

Ising, E.

[I1]   Beitrag zur Theorie des Ferromagnetismus, *Z. Physik* **31**, 253–258 (1925).

Jevons, W. S.

[J1]   "The Principles of Science," 2nd ed., pp. 134–146. MacMillan, London and New York, 1892.

[K1]   Graph theory and crystal physics, *in* "Graph Theory and Theoretical Physics" (F. Harary ed.), pp. 44–110. Academic Press, New York, 1967.

Katz, L.

[K2]   Probability of indecomposability of a random mapping function, *Ann. Math. Statist.* **26**, 512–517 (1955).

King, C.

[KP1]   Calculation of the number of graphs of order $p = 1$ (1) 24, to be published.
          See [HKMR1].

Kirchhoff, G.

[K3]   Über die Auflösung der Gleichungen, auf welche man bei der Untersuchung der linearen Verteilung galvanischer Ströme gefuhrt wird, *Ann. Phys. Chem.* **72**, 497–508 (1847).

Klarner, D. A.

[K4]   Cell growth problems, *Canad. J. Math.* **19**, 851–863 (1967).

Kraitchik, M.

[K5]   "Mathematical Recreations." Norton, New York, 1942.

Lawes, C. P.
          See Exercise 5.2.

Lynn, M. S.
          See [EHL1].

Mallows, C. L.

[MR1]   (with J. Riordan). The inversion enumerator for labeled trees, *Bull. Amer. Math. Soc.* **74**, 92–94 (1968).

Mayberry, J. P.
          See [BM2].

Moon, J.

[M1]   Various proofs of Cayley's formula for counting trees, *in* "A Seminar on Graph Theory" (F. Harary, ed.), pp. 70–78, Holt, New York, 1967.

[M2]   "Topics on Tournaments." Holt, New York, 1968.

[M3]   Connected graphs with unlabeled end-points, *J. Combinatorial Theory* **6**, 65–66 (1969).

[M4]   "Counting Labeled Trees." Canad. Math. Congress, Montreal, 1970.

[M5]   Almost all graphs have a spanning cycle, *Canad. Math. Bull.* **15**, 39–41 (1972).

[MM1]   (with L. Moser). Almost all tournaments are irreducible, *Canad. Math. Bull.* **5**, 61–65 (1962).

[MM2]   (with L. Moser). Almost all (0, 1) matrices are primitive, *Studia Sci. Math. Hungar.* **1**, 153–156 (1966).

      See also [BM1].

Mowshowitz, A.

      See [HMR1] and [HKMR1].

Norman, R. Z.

[N1]   On the number of linear graphs with given blocks, Dissertation, Univ. of Michigan, Ann Arbor, 1954.

      See also [FNU1], [HN1], [HN2], [HN3], and [HNC1].

Oberschelp, W.

[O1]   Kombinatorische Anzahlbestimmungen in Relationen, *Math. Ann.* **174**, 53–58 (1967).

Onsager, L.

[O2]   Crystal statistics I, a two-dimensional model with an order-disorder transition, *Phys. Rev.* **65**, 117–149 (1944).

Ore, O.

[OS1]   (with G. J. Stemple). Numerical methods in the four color problem, *in* "Recent Progress in Combinatorics" (W. T. Tutte, ed.). Academic Press, New York, 1969.

Osterweil, L.

[O3]   Enumeration of nonseparable graphs on fewer than 10 points, to be published.

Otter, R.

[O4]   The number of trees, *Ann. of Math.* **49**, 583–599 (1948).

Palmer, E. M.

[P1]   On the number of labeled 2-trees, *J. Combinatorial Theory* **6**, 206–207 (1969).

[P2]   Combinatorial problems in set-theoretic form, *Math. Mag.* **42**, 32–37 (1969).

[P3]   The exponentiation group as the automorphism group of a graph, *in* "Proof Techniques in Graph Theory" (F. Harary, ed.), pp. 125–131. Academic Press, New York, 1969.

[P4]   Methods for the enumeration of multigraphs, *in* "The Many Facets of Graph Theory" (G. Chartrand and S. Kapoor, eds.), pp. 251–261. Springer Publ., New York, 1969.

[P5]   Asymptotic formulas for the number of self-complementary graphs and digraphs, *Mathematika* **17**, 85–90 (1970).

[P6]   Graphical enumeration methods, *in* "New Directions in the Theory of Graphs." (F. Harary, ed.), Academic Press, New York, 1972.

[PR1]   (with R. C. Read). On the number of plane 2-trees *J. London Math. Soc.* to be published.

[PR2]   (with R. W. Robinson). The matrix group of two permutation groups, *Bull. Amer. Math. Soc.* **73**, 204–207 (1967).

[PR3]   (with R. W. Robinson). Enumeration under two representations of the wreath product, to be published.

      See also [HP1], [HP2], [HP3], [HP4], [HP5], [HP6], [HP7], [HP8], [HP9], [HP10], [HP11], [HP12], [HPR1], and [KP1].

Parthasarathy, K. R.
[P7]   Enumeration of ordinary graphs with given partition, *Canad. J. Math.* **20**, 40–47 (1968).

Pippert, R. E.
     See [BP1].

Pólya, G.
[P8]   Kombinatorische Anzahlbestimmungen für Gruppen, Graphen und chemische Verbindungen, *Acta Math.* **68**, 145–254 (1937).
[P9]   Sur les types des propositions composées, *J. Symbolic Logic* **5**, 98–103 (1940).

Prins, G.
     See [HP14], [HP15], and [HPT1].

Prüfer, H.
[P10]   Neuer Beweis eines Satzes über Permutationen, *Arch. Math. Phys.* **27**, 742–744 (1918).

Read, R. C.
[R1]   The enumeration of locally restricted graphs, I and II, *J. London Math. Soc.* **34**, 417–436 (1959); **35**, 344–351 (1960).
[R2]   The number of k-colored graphs on labelled nodes, *Canad. J. Math.* **12**, 409–413 (1960).
[R3]   A note on the number of functional digraphs, *Math. Ann.* **143**, 109–110 (1961).
[R4]   Contributions to the cell growth problem, *Canad. J. Math.* **14**, 1–20 (1962).
[R5]   Euler graphs on labelled nodes, *Canad. J. Math.* **14**, 482–486 (1962).
[R6]   On the number of self-complementary graphs and digraphs, *J. London Math. Soc.* **38**, 99–104 (1963).
[R7]   Some applications of a theorem of deBruijn, *in* "Graph Theory and Theoretical Physics" (F. Harary, ed.), pp. 273–280. Academic Press, New York, 1967.
[R8]   The use of S-functions in combinatory analysis, *Canad. J. Math.* **20**, 808–841 (1968).
[R9]   Some unusual enumeration problems, *Ann. N.Y. Acad. Sci.* **175**, 314–326 (1970).
[RW1]   (with E. M. Wright). Coloured graphs: A correction and extension, *Canad. J. Math.* to be published.
     See also [HKMR1], [HPR1], [PR1].

Redfield, J. H.
[R10]   The theory of group-reduced distributions, *Amer. J. Math.* **49**, 433–455 (1927).

Reidemeister, K.
[R11]   "Knotentheorie." Springer-Verlag, Berlin, 1932. Reprinted by Chelsea, New York, 1948.

Rényi, A.
[R12]   Some remarks on the theory of trees, *Publ. Math. Inst. Hungar. Acad. Sci.* **4**, 73–85 (1959).
[R13]   On connected graphs, *Publ. Math. Inst. Hungar. Acad. Sci.* **4**, 385–388 (1959).
     See also [ER1].

Riddell, R. J.
[R14]   Contributions to the theory of condensation, Dissertation, Univ. of Michigan, Ann Arbor, 1951.

Riordan, J.
[R15]   "An Introduction to Combinatorial Analysis." Wiley, New York, 1958.
[R16]   The enumeration of trees by height and diameter, *IBM J. Res. Develop.* **4**, 473–478 (1960).
     See also [AFPR1], [CR1], [GR1], and [HMR1].

Roberts, F. S.
[RS1]   (with J. H. Spencer). A characterization of clique graphs, *J. Combinatorial Theory* **10B**, 102–108 (1971).

Robinson, R. W.
[R17]   Enumeration of colored graphs, *J. Combinatorial Theory* **4**, 181–190 (1968).
[R18]   Enumeration of euler graphs, *in* "Proof Techniques in Graph Theory" (F. Harary, ed.), pp. 147–153. Academic Press, New York, 1969.
[R19]   Enumeration of non-separable graphs, *J. Combinatorial Theory* **9**, 327–356 (1970).
[R20]   Enumeration of acyclic digraphs, *in* "Combinatorial Mathematics and its Applications" (R. C. Bose *et al.* eds.), pp. 391–399. Univ. of North Carolina, Chapel Hill, 1970.
         See also [PR2] and [PR3].

Rudvalis, A.
[RS2]   (with E. Snapper). Numerical polynomials for arbitrary characters, *J. Combinatorial Theory* **10A**, 145–159 (1971).

Russell, B.
         See [WR1].

Scoins, H. I.
[S1]   The number of trees with nodes of alternate parity, *Proc. Cambridge Philos. Soc.* **58**, 12–16 (1962).

Sharp, Jr., H.
[S2]   Enumeration of vacuously transitive relations, *Discrete Math.* to be published.

Slepian, D.
[S3]   On the number of symmetry types of boolean functions of n variables, *Canad. J. Math.* **5**, 185–193 (1953).

Sloane, N. J.
[S4]   "Handbook of Integer Sequences," Academic Press, New York, 1973.

Smith, C. A. B.
[ST1]   (with W. T. Tutte). On unicursal paths in a network of degree 4, *Amer. Math. Monthly* **48**, 233–237 (1941).
         See also [BSST1].

Snapper, E.
         See [RS3].

Spencer, J. H.
         See [RS1].

Sridharan, M. R.
[S5]   Self-complementary and self-converse oriented graphs, *Indag. Math.* **32**, 441–447 (1970).

Stein, P.
         See Appendix I.

Stemple, G. J.
         See [OS1].

Stockmeyer, P. K.
[S6]   The enumeration of graphs with prescribed automorphism group, Dissertation, Univ. of Michigan, Ann Arbor, 1971.

Stone, A. H.
         See [BSST1].

Turner, J.
[T1]   Point-symmetric graphs with a prime number of points, *J. Combinatorial Theory* **3**, 136–145 (1967).

Tutte, W. T.
[T2]   The dissection of equilateral triangles into equilateral triangles, *Proc. Cambridge Philos. Soc.* **44**, 463–482 (1948).
[T3]   The number of planted trees with a given partition, *Amer. Math. Monthly* **71**, 272–277 (1964).
[T4]   The enumerative theory of planar maps, *in* "A Survey of Combinatorial Theory," (J. N. Srivastava *et al.* eds.) North-Holland, Amsterdam, 1972, to be published. See also [HPT1], [HT1], and [ST1].

Uhlenbeck, G. E.
[UF1]   (with G. W. Ford). "Lectures in Statistical Mechanics." *Amer. Math. Soc.*, Providence, Rhode Island, 1963. See also [FNU1], [FU1], and [HU1].

Walkup, D. W.
[W1]   The number of plane trees, *Mathematika*, to be published.

Whitehead, A. N.
[WR1]   (with B. Russell). "Principia Mathematica," Vol. II, p. 301. Cambridge Univ. Press London and New York, 1912.

Whitney, H.
[W2]   Congruent graphs and the connectivity of graphs, *Amer. J. Math.* **54** 150–168 (1932).

Wright, E. M.
[W3]   Counting coloured graphs, *Canad. J. Math.* **13**, 683–693 (1961).
[W4]   A relationship between two sequences I, II and III, *Proc. London Math. Soc.* **17**, 296–304, 547–552 (1967); *J. London Math. Soc.* **43**, 720–724 (1968).
[W5]   The number of graphs on many unlabelled nodes, *Math. Ann.* **183**, 250–253 (1969).
[W6]   Asymptotic enumeration of connected graphs, *Proc. Roy. Soc. Edinburgh Sect. A* **68**, 298–308 (1970). See also [RW1].
[W7]   Graphs on unlabelled nodes with a given number of edges, *Acta Math.* **168**, 1–9 (1971).

'Twas brillig and the slithy toves
Did gyre and gimble in the wabe
All mimsy were the borogoves
And the mome raths outgrabe.

Lewis Carroll, *Jabberwocky*

INDEX OF SYMBOLS

The symbols, letters and notational devices as they are most commonly used in this book are collected here. They are arranged in four categories: Latin letters, Greek letters, script letters, and operations on graphs.

It isn't really
Anywhere!
It's somewhere else
Instead!

A. A. Milne

INDEX OF DEFINITIONS

We have included in the text the definitions of almost all concepts used in this book. For the convenience of the reader, a few definitions even appear more than once. They are all listed in this index. The terminology follows the book on graph theory [H1] which provides a useful supplement.